Bioinstrumentation

Bioinstrumentation

John D. Enderle

ISBN-13: 978-3-031-00488-9   paperback
ISBN-13: 978-3-031-01616-5   ebook

DOI 10.1007/978-3-031-01616-5

A Publication in the Springer series

SYNTHESIS LECTURES ON BIOMEDICAL ENGINEERING
Lecture #6
Series Editor and Affiliation: John D. Enderle, University of Connecticut

1930-0328   Print
1930-0336   Electronic

First Edition
10 9 8 7 6 5 4 3 2 1

Portions of this manuscript were reprinted from the following book with the Permission of Elsevier, *INTRODUCTION TO BIOMEDICAL ENGINEERING, ISBN 0122386626 2005, Chapter 8, pp403–504, Enderle et al 2nd edition*. This permission is granted for non-exclusive world English rights only in both print and on the world wide web.

Additional Information regarding the bestselling book, *Introduction to Biomedical Engineering*, 2nd edition 2005, by Enderle, Blanchard and Bronzino, from Elsevier can be found on the Elsevier Homepage http://www/elsevier.com.

Bioinstrumentation

John D. Enderle
Program Director & Professor for Biomedical Engineering
University of Connecticut

SYNTHESIS LECTURES ON BIOMEDICAL ENGINEERING #6

ABSTRACT

This short book provides basic information about bioinstrumentation and electric circuit theory. Many biomedical instruments use a transducer or sensor to convert a signal created by the body into an electric signal. Our goal here is to develop expertise in electric circuit theory applied to bioinstrumentation. We begin with a description of variables used in circuit theory, charge, current, voltage, power and energy. Next, Kirchhoff's current and voltage laws are introduced, followed by resistance, simplifications of resistive circuits and voltage and current calculations. Circuit analysis techniques are then presented, followed by inductance and capacitance, and solutions of circuits using the differential equation method. Finally, the operational amplifier and time varying signals are introduced. This lecture is written for a student or researcher or engineer who has completed the first two years of an engineering program (i.e., 3 semesters of calculus and differential equations). A considerable effort has been made to develop the theory in a logical manner—developing special mathematical skills as needed. At the end of the short book is a wide selection of problems, ranging from simple to complex.

KEYWORDS

Bioinstrumentation, Circuit Theory, Introductory Biomedical Engineering, Sensors, Transducers, Circuits, Voltage, Current

Contents

Preface

This short book on bioinstrumentation is written for a reader who has completed the first two years of an engineering program (i.e., three semesters of calculus and differential equations). A considerable effort has been made to develop the theory in a logical manner—developing special mathematical skills as needed.

I have found it best to introduce this material using simple examples followed by more difficult ones.

At the end of the short book is a wide selection of problems, ranging from simple to difficult, presented in the same general order as covered in the textbook.

I acknowledge and thank William Pruehsner for the technical illustrations. Portions of this short book are from Chapter 8 of Enderle, J. D., Blanchard, S. M., and Bronzino, J. D., *Introduction to Biomedical Engineering (Second Edition)*, Elsevier, Amsterdam, 2005, 1118 pages, with Sections 1, 2 and 13 contributed by Susan Blanchard, Amanda Marley, and H. Troy Nagle.

CHAPTER 1

Introduction

This short book provides basic information about bioinstrumentation and electric circuit theory. Many biomedical instruments use a transducer or sensor to convert a signal created by the body into an electric signal. Our goal here is to develop expertise in electric circuit theory applied to bioinstrumentation. We begin with a description of variables used in circuit theory, charge, current, voltage, power, and energy. Next, Kirchhoff's current and voltage laws are introduced, followed by resistance, simplifications of resistive circuits and voltage and current calculations. Circuit analysis techniques are then presented, followed by inductance and capacitance, and solutions of circuits using the differential equation method. Finally, the operational amplifier and time-varying signals are introduced.

Before 1900, medicine had little to offer the typical citizen because its resources were mainly the education and little black bag of the physician. The origins of the changes that occurred within medical science are found in several developments that took place in the applied sciences. During the early 19th century, diagnosis was based on physical examination, and treatment was designed to heal the structural abnormality. By the late 19th century, diagnosis was based on laboratory tests, and treatment was designed to remove the cause of the disorder. The trend towards the use of technology accelerated throughout the 20th century. During this period, hospitals became institutions of research and technology. Professionals in the areas of chemistry, physics, mechanical engineering, and electrical engineering began to work in conjunction with the medical field, and biomedical engineering became a recognized profession. As a result, medical technology advanced more in the 20th century than it had in the rest of history combined (Fig. 1.1).

During this period, the area of electronics had a significant impact on the development of new medical technology. Men such as Richard Caton and Augustus Desire proved that the human brain and heart depended upon bioelectric events. In 1903, William Einthoven expanded on these ideas after he created the first string galvanometer. Einthoven placed two skin sensors on a man and attached them to the ends of a silvered wire that was suspended through holes drilled in both ends of a large permanent magnet. The suspended silvered wire moved rhythmically as the subject's heart beat. By projecting a tiny light beam across the silvered

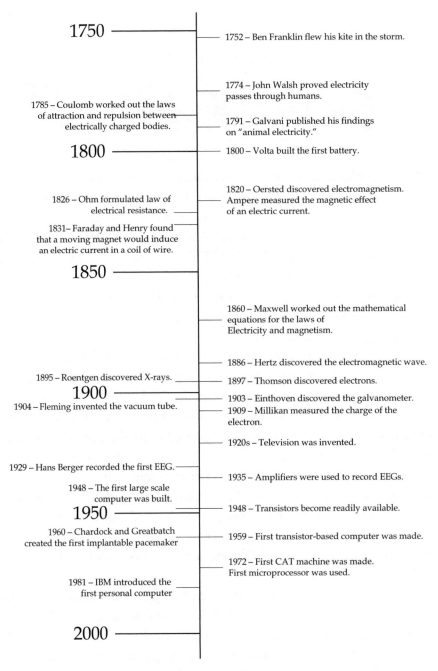

FIGURE 1.1: Timeline for major inventions and discoveries that led to modern medical instrumentation.

wire, Einthoven was able to record the movement of the wire as waves on a scroll of moving photographic paper. Thus, the invention of the string galvanometer led to the creation of the electrocardiogram (ECG), which is routinely used today to measure and record the electrical activity of abnormal hearts and to compare those signals to normal ones.

In 1929, Hans Berger created the first electroencephalogram (EEG), which is used to measure and record electrical activity of the brain. In 1935, electrical amplifiers were used to prove that the electrical activity of the cortex had a specific rhythm, and, in 1960, electrical amplifiers were used in devices such as the first implantable pacemaker that was created by William Chardack and Wilson Greatbatch. These are just a small sample of the many examples in which the field of electronics has been used to significantly advance medical technology.

Many other advancements that were made in medical technology originated from research in basic and applied physics. In 1895, the X-ray machine, one of the most important technological inventions in the medical field, was created when W. K. Roentgen found that X-rays could be used to give pictures of the internal structures of the body. Thus, the X-ray machine was the first imaging device to be created.

Another important addition to medical technology was provided by the invention of the computer, which allowed much faster and more complicated analyses and functions to be performed. One of the first computer-based instruments in the field of medicine, the sequential multiple analyzer plus computer, was used to store a vast amount of data pertaining to clinical laboratory information. The invention of the computer made it possible for laboratory tests to be performed and analyzed faster and more accurately.

The first large-scale computer-based medical instrument was created in 1972 when the computerized axial tomography (CAT) machine was invented. The CAT machine created an image that showed all of the internal structures that lie in a single plane of the body. This new type of image made it possible to have more accurate and easier diagnosis of tumors, hemorrhages, and other internal damage from information that was obtained noninvasively.

Telemedicine, which uses computer technology to transmit information from one medical site to another, is being explored to permit access to health care for patients in remote locations. Telemedicine can be used to let a specialist in a major hospital receive information on a patient in a rural area and send back a plan of treatment specific for that patient.

Today, there is a wide variety of medical devices and instrumentation systems. Some are used to monitor patient conditions or acquire information for diagnostic purposes, e.g. ECG and EEG machines, while others are used to control physiological functions, e.g. pacemakers and ventilators. Some devices, like pacemakers, are implantable while many others are used noninvasively. This chapter will focus on those features that are common to devices that are used to acquire and process physiological data.

CHAPTER 2

Basic Bioinstrumentation System

The quantity, property, or condition that is measured by an instrumentation system is called the measurand (Fig. 2.1). This can be a bioelectric signal, such as those generated by muscles or the brain, or a chemical or mechanical signal that is converted to an electrical signal. Sensors are used to convert physical measurands into electric outputs. The outputs from these biosensors are analog signals, i.e. continuous signals, which are sent to the analog processing and digital conversion block. There, the signals are amplified, filtered, conditioned, and converted to digital form. Methods for modifying analog signals, such as amplifying and filtering an ECG signal, are discussed later in this chapter. Once the analog signals have been digitized and converted to a form that can be stored and processed by digital computers, many more methods of signal conditioning can be applied.

Basic instrumentation systems also include output display devices that enable human operators to view the signal in a format that is easy to understand. These displays may be numerical or graphical, discrete or continuous, and permanent or temporary. Most output display devices are intended to be observed visually, but some also provide audible output, e.g. a beeping sound with each heart beat.

In addition to displaying data, many instrumentation systems have the capability of storing data. In some devices, the signal is stored briefly so that further processing can take place or so that an operator can examine the data. In other cases, the signals are stored permanently so that different signal processing schemes can be applied at a later time. Holter monitors, for example, acquire 24 hrs of ECG data that is later processed to determine arrhythmic activity and other important diagnostic characteristics.

With the invention of the telephone and now with the Internet, signals can be acquired with a device in one location, perhaps in a patient's home, and transmitted to another device for processing and/or storage. This has made it possible, for example, to provide quick diagnostic feedback if a patient has an unusual heart rhythm while at home. It has also allowed medical facilities in rural areas to transmit diagnostic images to tertiary care hospitals so that specialized physicians can help general practitioners arrive at more accurate diagnoses.

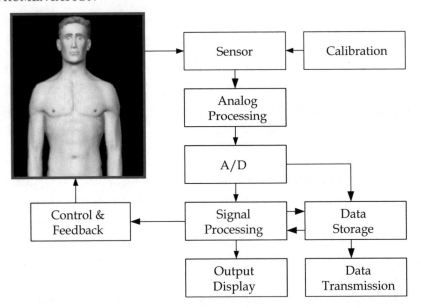

FIGURE 2.1: Basic instrumentation systems using sensors to measure a signal with data acquisition, storage and display capabilities, along with control and feedback.

Two other components play important roles in instrumentation systems. The first is the calibration signal. A signal with known amplitude and frequency content is applied to the instrumentation system at the sensor's input. The calibration signal allows the components of the system to be adjusted so that the output and input have a known, measured relationship. Without this information, it is impossible to convert the output of an instrument system into a meaningful representation of the measurand.

Another important component, a feedback element, is not a part of all instrumentation systems. These devices include pacemakers and ventilators that stimulate the heart or the lungs. Some feedback devices collect physiological data and stimulate a response, e.g. a heart beat or breath, when needed or are part of biofeedback systems in which the patient is made aware of a physiological measurement, e.g. blood pressure, and uses conscious control to change the physiological response.

CHAPTER 3

Charge, Current, Voltage, Power and Energy

3.1 CHARGE

Two kinds of charge, positive and negative, are carried by protons and electrons, respectively. The negative charge carried by an electron, q_e, is the smallest amount of charge that exists and is measured in units called Coulombs (C).

$$q_e = -1.602 \times 10^{-19}\,\text{C}$$

The symbol $q(t)$ is used to represent charge that changes with time, and Q for constant charge. The charge carried by a proton is the opposite of the electron.

Example 3.1. What is the charge of 3×10^{-15}g of electrons?

Solution

$$Q = 3 \times 10^{-15}\text{g} \times \frac{1\,\text{kg}}{10^3\,\text{g}} \times \frac{1\ \text{electron}}{9.1095 \times 10^{-31}\,\text{kg}} \times \frac{-1.60219 \times 10^{-19}\text{C}}{\text{electron}}$$

$$= -5.27643 \times 10^{-7}\,\text{C}$$

∎

3.2 CURRENT

Electric current, $i(t)$, is defined as the change in the amount of charge that passes through a given point or area in a specified time period. Current is measured in amperes (A). By definition, one ampere equals one coulomb/second (C/s).

$$i(t) = \frac{dq}{dt} \tag{3.1}$$

and

$$q(t) = \int_{t_0}^{t} i(\lambda)\,d\lambda + q(t_0) \tag{3.2}$$

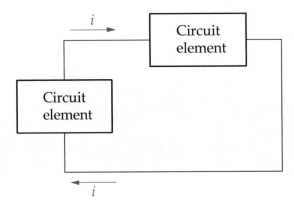

FIGURE 3.1: A simple electric circuit illustrating current flowing around a closed loop.

Current, defined by Eq. (3.1), also depends on the direction of flow as illustrated in the circuit in Fig. 3.1.

Example 3.2. Suppose the following current is flowing in Fig. 3.1.

$$i(t) = \begin{cases} 0 & t < 0 \\ 3e^{-100t}\,A & t \geq 0 \end{cases}$$

Find the total charge.

Solution

$$Q = \int_{-\infty}^{\infty} i\,dt = \int_{0}^{\infty} 3e^{-100t}\,dt = -\frac{3}{100}e^{-100t}\Big|_{t=0}^{\infty} = 0.03\ \text{C}$$ ∎

Consider Fig. 3.1. Current is defined as positive if

    (a) A positive charge is moving in the direction of the arrow

    (b) A negative charge is moving in the opposite direction of the arrow

Since these two possibilities produce the same outcome, there is no need to be concerned as to which is responsible for the current. In electric circuits, current is carried by electrons in metallic conductors.

Current is typically a function of time, as given by Eq. (3.1). Consider Fig. 3.2 with the current entering terminal 1 in the circuit on the right. In the time interval 0–1.5 s, current is positive and enters terminal 1. In the time interval 1.5–3 s, the current is negative and enters terminal 2 with a positive value. We typically refer to a constant current as a DC current, and

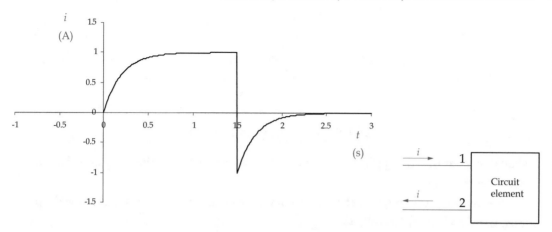

FIGURE 3.2: (Left) A sample current waveform. (Right) A circuit element with current entering terminal 1 and leaving terminal 2. Passive circuit elements have two terminals with a known voltage–current relationship. Examples of passive circuit elements include resistors, capacitors and inductors.

denote it with a capital letter such as I indicating it does not change with time. We denote a time-varying current with a lower case letter, such as $i(t)$, or just i.

3.2.1 Kirchhoff's Current Law

Current can flow only in a closed circuit, as shown in Fig. 3.1. No current is lost as it flows around the circuit because net charge cannot accumulate within a circuit element and charge must be conserved. Whatever current enters one terminal must leave at the other terminal. Since charge cannot be created and must be conserved, the sum of the currents at any node, that is, a point at which two or more circuit elements have a common connection, must equal zero so no net charge accumulates. This principle is known as Kirchhoff's current law (KCL), given as

$$\sum_{n=1}^{N} i_n(t) = 0 \qquad (3.3)$$

where there are N currents leaving the node. Consider the circuit in Fig. 3.3. Using Eq. (3.3) and applying KCL for the currents *leaving* the node gives

$$-i_1 - i_2 + i_4 + i_3 = 0$$

The previous equation is equivalently written for the currents *entering* the node, as

$$i_1 + i_2 - i_4 - i_3 = 0$$

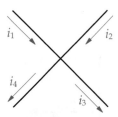

FIGURE 3.3: A node with 4 currents.

It should be clear that the application of KCL is for *all* currents, whether they are all leaving or all entering the node.

In describing a circuit, we define its characteristics with the terms "node", "branch", "path", "closed path" and "mesh" as follows.

- *Node:* A point at which two or more circuit elements have a common connection.
- *Branch:* A circuit element or connected group of circuit elements. A connected group of circuit elements usually connect nodes together.
- *Path*: A connected group of circuit elements in which none is repeated.
- *Closed Path*: A path that starts and ends at the same node.
- *Mesh*: A closed path that does not contain any other closed paths within it.
- *Essential Node*: A point at which three or more circuit elements have a common connection.
- *Essential Branch*: A branch that connects two essential nodes.

In Fig. 3.4, there are five nodes, A, B, C, D and E, which are all essential nodes. Kirchhoff's current law is applied to each of the nodes as follows.

$$\text{Node A:} \quad -i_1 + i_2 - i_3 = 0$$
$$\text{Node B:} \quad i_3 + i_4 + i_5 - i_6 = 0$$
$$\text{Node C:} \quad i_1 - i_4 - i_8 = 0$$
$$\text{Node D:} \quad -i_7 - i_5 + i_8 = 0$$
$$\text{Node E:} \quad -i_2 + i_6 + i_7 = 0$$

Kirchhoff's current law is also applicable to any closed surface surrounding a part of the circuit. It is understood that the closed surface does not intersect any of the circuit elements. Consider the closed surface drawn with dashed lines in Fig. 3.4. Kirchhoff's current law applied to the closed surface gives

$$-i_1 + i_4 + i_5 + i_7 = 0$$

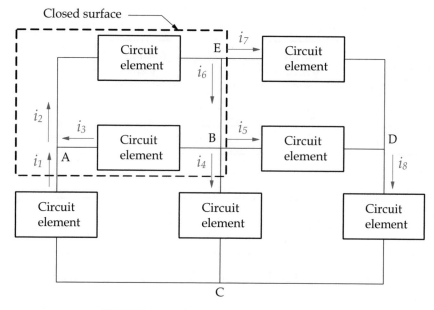

FIGURE 3.4: A circuit with a closed surface.

3.3   VOLTAGE

Voltage represents the work per unit charge associated with moving a charge between two points (A and B in Fig. 3.5), and given as

$$v = \frac{dw}{dq} \qquad (3.4)$$

The unit of measurement for voltage is the volt (V). A constant (DC) voltage source is denoted by V while a time-varying voltage is denoted by $v(t)$, or just v. In Fig. 3.5, the voltage, v, between two points (A and B), is the amount of energy required to move a charge from point A to point B.

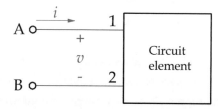

FIGURE 3.5: Voltage and current convention.

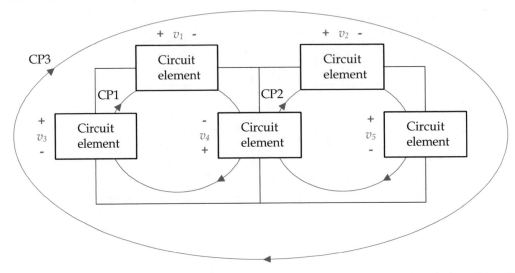

FIGURE 3.6: Circuit illustrating Kirchhoff's voltage law. Closed paths are identified as CP1, CP2 and CP3.

3.3.1   Kirchhoff's Voltage Law

Kirchhoff's voltage law (KVL) states the sum of all voltages in a closed path is zero, or

$$\sum_{n=1}^{N} v_n(t) = 0 \qquad (3.5)$$

where there are N voltage drops assigned around the closed path, with $v_n(t)$ denoting the individual voltage drops. The sign for each voltage drop in Eq. (3.5) is the first sign encountered while moving around the closed path.

Consider the circuit in Fig. 3.6, with each circuit element assigned a voltage, v_n, with a given polarity, and three closed paths, CP1, CP2 and CP3. Kirchhoff's voltage law for each closed path is given as

$$\text{CP1:} \quad -v_3 + v_1 - v_4 = 0$$
$$\text{CP2:} \quad v_4 + v_2 + v_5 = 0$$
$$\text{CP3:} \quad -v_3 + v_1 + v_2 + v_5 = 0$$

Kirchhoff's laws are applied in electric circuit analysis to determine unknown voltages and currents. Each unknown variable has its distinct equation. To solve for the unknowns using MATLAB, we create a matrix representation of the set of equations and solve. This method is demonstrated in many examples in this book.

Example 3.3. Find I_1, I_2 and I_3 for the following circuit.

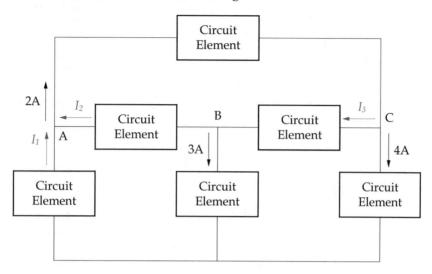

Solution. We apply KCL first at node C, then B and finally A.

$$\text{Node C: } -2 + I_3 + 4 = 0; \quad I_3 = -2 \text{ A}$$

$$\text{Node B: } I_2 + 3 - I_3 = 0; \quad I_2 = I_3 - 3 = -5 \text{ A}$$

$$\text{Node A: } -I_1 - I_2 + 2 = 0; \quad I_1 = 2 - I_2 = 7 \text{ A}$$ ∎

3.4   POWER AND ENERGY

Power is the rate of energy expenditure given as

$$p = \frac{dw}{dt} = \frac{dw}{dq}\frac{dq}{dt} = vi \tag{3.6}$$

where p is power measured in watts (W), and w is energy measured in joules (J). Power is usually determined by the product of voltage across a circuit element and the current through it. By convention, we assume that a positive value for power indicates that power is being delivered (or absorbed or consumed) by the circuit element. A negative value for power indicates that power is being extracted or generated by the circuit element, i.e., a battery.

Figure 3.7 illustrates the four possible cases for a circuit element's current and voltage configuration. According to convention, if both i and v are positive, with the arrow and polarity shown in Fig. 3.7A, energy is absorbed (either lost by heat or stored). If either the current arrow or voltage polarity is reversed as in B and C, energy is supplied to the circuit. Note that if both the current direction and voltage polarity are reversed together, as in D, energy is absorbed.

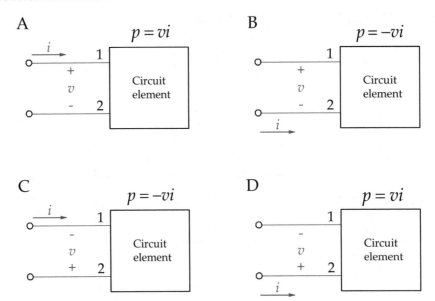

FIGURE 3.7: Polarity references for four cases of current and voltage. Cases A and D result in positive power being consumed by the circuit element. Cases B and C result in negative power being extracted from the circuit element.

A passive circuit element is defined as an element whose power is always positive or zero, which may be dissipated as heat (resistance), stored in an electric field (capacitor) or stored in a magnetic field (inductor). We define an active circuit element as one whose power is negative and capable of generating energy.

Energy is given by

$$w(t) = \int_{-\infty}^{t} p\,dt \qquad (3.7)$$

3.5   SOURCES

Sources are two terminal devices that provide energy to a circuit. There is no direct voltage–current relationship for a source; when one of the two variables is given, the other cannot be determined without knowledge of the rest of the circuit. Independent sources are devices for which the voltage or current is given and the device maintains its value regardless of the rest of the circuit. A device that generates a prescribed voltage at its terminals, regardless of the current flow, is called an ideal voltage source. Figure 3.8A and B shows the general symbols for an ideal voltage source. Figure 3.8C shows an ideal current source that delivers a prescribed current to

FIGURE 3.8: Basic symbols used for independent sources. (A) Battery. (B) Ideal voltage source. V_s can be a constant DC source (Battery) or a time-varying source. (C) Ideal current source I_s.

FIGURE 3.9: Basic symbols used for dependent or controlled sources. (Left) Controlled voltage source. The voltage V_s is a known function of some other voltage or current in the circuit. (Right) Controlled current source. The current I_s is a known function of some other voltage or current in the circuit.

the attached circuit. The voltage generated by an ideal current source depends on the elements in the rest of the circuit.

Shown in Fig. 3.9 are a dependent voltage and current source. A dependent source takes on a value equaling a known function of some other voltage or current value in the circuit. We use a diamond-shaped symbol to represent a dependent source. Often, a dependent source is called a controlled source. The current generated for a dependent voltage source and the voltage for a dependent current source depend on circuit elements in the rest of the circuit. Dependent sources are very important in electronics. Later in this chapter, we will see that the operational amplifier uses a controlled voltage source for its operation.

Example 3.4. Find the voltage V_3 for the circuit shown in the following figure.

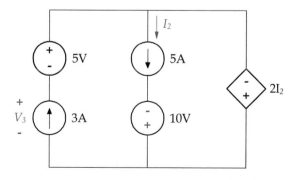

Solution.   Current I_2 equals 5 A because it is in the same branch as the 5 A current source. The voltage across the dependent voltage source on the right side of the circuit equals $2I_2 = 10$ V. Applying KVL around the outer closed path gives

$$-V_3 - 5 - 10 = 0$$

or

$$V_3 = -15 \text{ V} \qquad\qquad\qquad \blacksquare$$

CHAPTER 4

Resistance

4.1 RESISTORS

A resistor is a circuit element that limits the flow of current through it and is denoted with the symbol ⋀⋁⋀. Resistors are made of different materials and their ability to impede current is given with a value of resistance, denoted R. Resistance is measured in Ohms (Ω), where $1\,\Omega = 1\,\mathrm{V/A}$. A theoretical bare wire that connects circuit elements together has a resistance of zero. A gap between circuit elements has a resistance of infinity. An ideal resistor follows Ohm's law, which describes a linear relationship between voltage and current, as shown in Fig. 4.1, with a slope equal to the resistance.

There are two ways to write Ohm's law, depending on the current direction and voltage polarity. Ohm's law is written for Fig. 4.2A as

$$v = iR \qquad\qquad (4.1)$$

and for Fig. 4.2B as

$$v = -iR \qquad\qquad (4.2)$$

In this book, we will use the convention shown in Fig. 4.2A to write the voltage drop across a resistor. As described, the voltage across a resistor is equal to the product of the current flowing through the element and its resistance, R. This linear relationship does not apply at very high voltages and currents. Some electrically conducting materials have a very small range of currents and voltages in which they exhibit linear behavior. This is true of many physiological models as well: linearity is observed only within a range of values. Outside this range, the model is nonlinear. We define a short circuit as shown in Fig. 4.3A with $R = 0$, and having a 0 V voltage drop. We define an open circuit as shown in Fig. 4.3B with $R = \infty$, and having 0 A current pass through it.

Each material has a property called resistivity (ρ) that indicates the resistance of the material. Conductivity (σ) is the inverse of resistivity, and conductance (G) is the inverse of

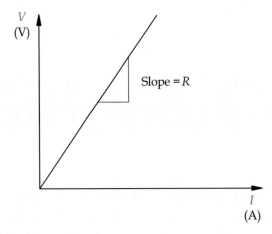

FIGURE 4.1: Voltage–current relationship for a resistor.

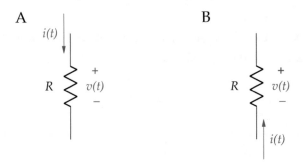

FIGURE 4.2: An ideal resister with resistance R in Ohms (Ω).

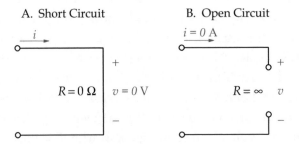

FIGURE 4.3: Short and open circuits.

resistance. Conductance is measured in units called siemens (S) and has units of A/V. In terms of conductance, Ohm's law is written as

$$i = Gv \qquad (4.3)$$

Example 4.1. From the following circuit, find I_2, I_3 and V_1.

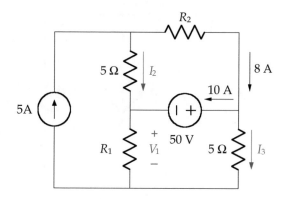

Solution. First we find I_2 by applying KCL at the node in the upper left of the circuit.

$$-5 + I_2 + 8 = 0$$

and

$$I_2 = -3\,\text{A}$$

Current I_3 is determined by applying KCL at the node on the right of the circuit.

$$10 + I_3 - 8 = 0$$

and

$$I_3 = -2\,\text{A}$$

Voltage V_1 is determined by applying KVL around the lower right closed path and using Ohm's law.

$$-V_1 - 50 + 5I_3 = 0$$
$$V_1 = -50 + 5 \times (-2) = -60\,\text{V} \qquad \blacksquare$$

4.2 POWER

The power consumed by a resistor is given by the combination of Eq. (3.6) and either Eq. (4.1) or (4.2) as

$$p = vi = \frac{v^2}{R} = i^2 R \qquad (4.4)$$

and given off as heat. Equation (4.4) demonstrates that regardless of the voltage polarity and current direction, power is consumed by a resistor. Power is always positive for a resistor, which is true for any passive element.

Example 4.2.   Calculate the power in each element.

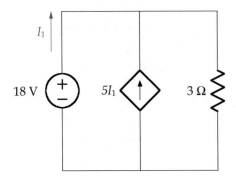

Solution. We first note that the resistor on the right has $18\,\text{V}$ across it, and therefore the current through it, according to Ohm's law, is $\frac{18}{3} = 6\,\text{A}$. To find the current I_1 we apply KCL at the upper node, giving

$$-I_1 - 5I_1 + 6 = 0$$

or

$$I_1 = 1\,\text{A}$$

The power for each of the circuit elements is

$$p_{18V} = -I_1 \times 18 = -18\,\text{W}$$
$$p_{5i_1} = -18 \times 5I_1 = -90\,\text{W}$$
$$p_{3\Omega} = \frac{18^2}{3} = 108\,\text{W}$$

In any circuit, the power supplied by the active elements always equals the power consumed. Here, the power generated is 108 W and the power consumed is 108 W, as required.   ∎

Example 4.3.   Electric safety is of paramount importance in a hospital or clinical environment. If sufficient current is allowed to flow through the body, significant damage can occur, as illustrated in the following figure.

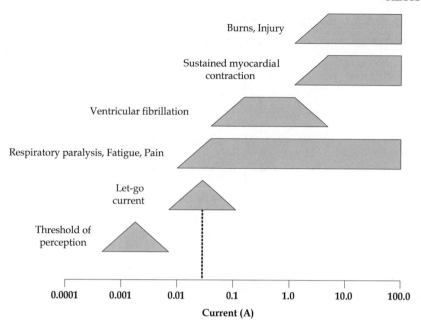

For instance, a current of magnitude 50 mA (dashed line) is enough to cause ventricular fibrillation, as well as other conditions. The figure on the left shows the current distribution from a macroshock from one arm to another (Redrawn from Enderle *et al.*, *Introduction to Biomedical Engineering, 2000*). A crude electric circuit model of the body consisting of two arms (each with resistance R_A), two legs (each with resistance R_L), body trunk (with resistance R_T), and head (with resistance R_H) is shown in the following figure on the right.

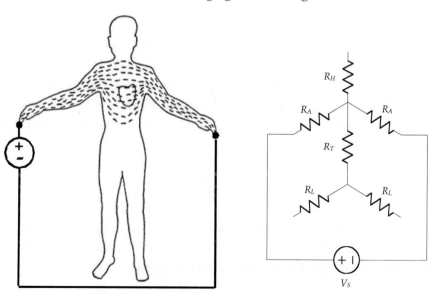

Since the only elements that form a closed path that current can flow is given by the source in series with the two arms, we reduce the body electric circuits to

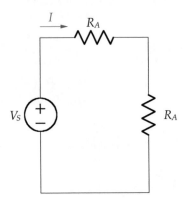

If $R_A = 400\,\Omega$ and $V_s = 120\,\text{V}$, then find I.

Solution. Using Ohm's law, we have

$$I = \frac{V_s}{R_A + R_A} = \frac{120}{800} = 0.15\,\text{A}$$

The current I is the current passing through the heart, and at this level it would cause ventricular fibrillation. ∎

4.3 EQUIVALENT RESISTANCE

It is sometimes possible to reduce complex circuits into simpler and, as we will see, equivalent circuits. We consider two circuits equivalent if they cannot be distinguished from each other by voltage and current measurements, that is, the two circuits behave identically. Consider the two circuits A and B in Fig. 4.4, consisting of combinations of resistors, each stimulated by a DC voltage V_s. These two circuits are equivalent if $I_A = I_B$. We represent the resistance of either circuit using Ohm's law as

$$R_{EQ} = \frac{V_s}{I_A} = \frac{V_s}{I_B} \tag{4.5}$$

Thus, it follows that any circuit consisting of resistances can be replaced by an equivalent circuit as shown in Fig. 4.5. In another section on a Thévenin equivalent circuit, we will expand this remark to include any combination of sources and resistances.

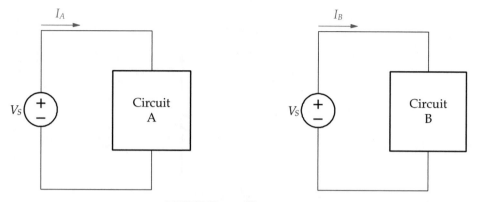

FIGURE 4.4: Two circuits.

4.4 SERIES AND PARALLEL COMBINATIONS OF RESISTANCE

4.4.1 Resistors in Series

If the same current flows from one resistor to another, the two are said to be in series. If these two resistors are connected to a third and the same current flows through all of them, then the three resistors are in series. In general, if the same current flows through N resistors, then the N resistors are in series. Consider Fig. 4.6 with three resistors in series. An equivalent circuit can be derived through KVL as

$$-V_s + IR_1 + IR_2 + IR_3 = 0$$

or rewritten in terms of an equivalent resistance R_{EQ} as

$$R_{EQ} = \frac{V_s}{I} = R_1 + R_2 + R_3$$

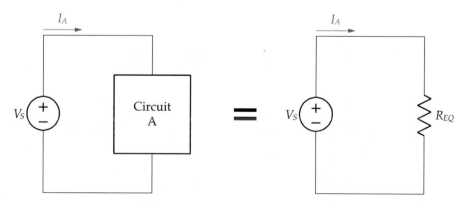

FIGURE 4.5: Equivalent circuits.

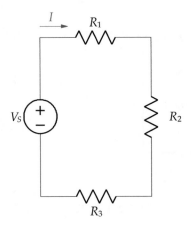

FIGURE 4.6: A series circuit.

In general, if we have N resistors in series,

$$R_{EQ} = \sum_{i=1}^{N} R_i \qquad (4.6)$$

4.4.2   Resistors in Parallel

Two or more elements are said to be in parallel if the same voltage is across each of the resistors. Consider the three parallel resistors as shown in Fig. 4.7. We use a shorthand notation to represent resistors in parallel using the $\|$ symbol. Thus in Fig. 4.7, $R_{EQ} = R_1 \| R_2 \| R_3$. An equivalent circuit for Fig. 4.7 is derived through KCL as

$$-I + \frac{V_s}{R_1} + \frac{V_s}{R_2} + \frac{V_s}{R_3} = 0$$

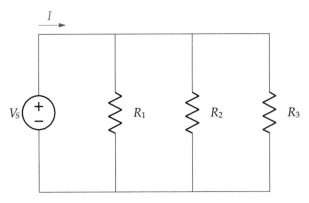

FIGURE 4.7: A parallel circuit.

or rewritten in terms of an equivalent resistance R_{EQ} as

$$R_{EQ} = \frac{V_s}{I} = \frac{1}{\frac{1}{R_1} + \frac{1}{R_2} + \frac{1}{R_3}}$$

In general, if we have N resistors in parallel,

$$R_{EQ} = \frac{1}{\frac{1}{R_1} + \frac{1}{R_2} + \cdots + \frac{1}{R_N}} \qquad (4.7)$$

For just two resistors in parallel, Eq. (4.7) is written as

$$R_{EQ} = R_1 \| R_2 = \frac{R_1 R_2}{R_1 + R_2} \qquad (4.8)$$

Example 4.4.   Find R_{EQ} and the power supplied by the source for the following circuit.

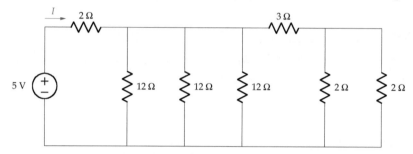

Solution.   To solve for R_{EQ}, apply from right to left the parallel and series combinations. First, we have two 2 Ω resistors in parallel that are in series with the 3 Ω resistor. Next, this group is in parallel with the three 12 Ω resistors. Finally, this group is in series with the 2 Ω resistor. These combinations are shown in the following figure and calculation:

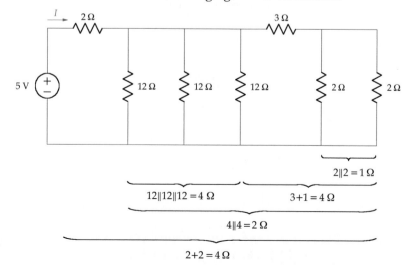

$$R_{EQ} = 2\,\Omega + ((12\,\Omega \parallel 12\,\Omega \parallel 12\,\Omega) \parallel (3\,\Omega + (2\,\Omega \parallel 2\,\Omega)))$$

$$= 2 + \left(\left(\frac{1}{\frac{1}{12} + \frac{1}{12} + \frac{1}{12}}\right) \parallel \left(3 + \frac{1}{\frac{1}{2} + \frac{1}{2}}\right)\right)$$

$$= 2 + ((4) \parallel (3 + 1)) = 2 + 2 = 4\,\Omega$$

Accordingly,

$$I = \frac{5}{R_{EQ}} = \frac{5}{4} = 1.25\,\text{A}$$

and

$$p = 5 \times I = 6.25\,\text{W} \qquad\qquad \blacksquare$$

4.5 VOLTAGE AND CURRENT DIVIDER RULES

Let us now extend the concept of equivalent resistance, $R_{EQ} = \frac{V}{I}$, to allow us to quickly calculate voltages in series resistor circuits and currents in parallel resistor circuits without digressing to the fundamentals.

4.5.1 Voltage Divider Rule

The voltage divider rule allows us to easily calculate the voltage across a given resistor in a series circuit. Consider finding V_2 in the series circuit shown in Fig. 4.8, where $R_{EQ} = R_1 + R_2$. Accordingly,

$$I = \frac{V_s}{R_{EQ}} = \frac{V_s}{R_1 + R_2}$$

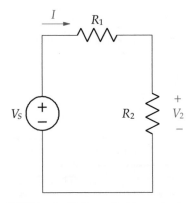

FIGURE 4.8: Voltage divider rule circuit.

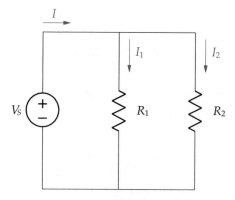

FIGURE 4.9: Current divider rule circuit.

and therefore

$$V_2 = IR_2 = V_s \frac{R_2}{R_1 + R_2}$$

This same analysis can be used to find V_1 as

$$V_1 = V_s \frac{R_1}{R_1 + R_2}$$

In general, if a circuit contains N resistors in series, the voltage divider rule gives the voltage across any one of the resistors, R_i, as

$$V_i = V_s \frac{R_i}{R_1 + R_2 + \cdots R_N} \tag{4.9}$$

4.5.2   Current Divider Rule

The current divider rule allows us to easily calculate the current through any resistor in parallel resistor circuits. Consider finding I_2 in the parallel circuit shown in Fig. 4.9, where $R_{EQ} = \frac{R_1 R_2}{R_1 + R_2}$. Accordingly,

$$I_2 = \frac{V_s}{R_2}$$

and

$$V_s = I \frac{R_1 R_2}{R_1 + R_2}$$

yielding after substituting V_s

$$I_2 = I \frac{\frac{1}{R_2}}{\frac{1}{R_1} + \frac{1}{R_2}}$$

In general, if a circuit contains N resistors in parallel, the current divider rule gives the current through any one of the resistors, R_i, as

$$I_i = I \frac{\frac{1}{R_i}}{\frac{1}{R_1} + \frac{1}{R_2} \cdots + \frac{1}{R_N}} \qquad (4.10)$$

Example 4.5. For the following circuit, find I_1.

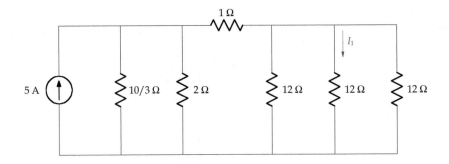

Solution. We solve this circuit problem in two parts, as is evident from the redrawn circuit that follows, by first finding I_2 and then I_1.

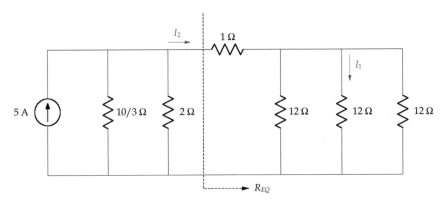

To begin, first find R_{EQ}, which, when placed into the circuit, reduces to three parallel resistors from which I_2 is calculated. The equivalent resistance is found as

$$R_{EQ} = 1 + (12 \,\|\, 12 \,\|\, 12) = 1 + \frac{1}{\frac{1}{12} + \frac{1}{12} + \frac{1}{12}} = 5\,\Omega$$

Applying the current divider rule on the three parallel resistors, $\frac{10}{3} \parallel 2 \parallel R_{EQ}$, we have

$$I_2 = 5 \left(\frac{\frac{1}{5}}{\frac{3}{10} + \frac{1}{2} + \frac{1}{5}} \right) = 1\,\text{A}$$

I_2 flows through the 1 Ω resistor, and then divides into three equal currents of $\frac{1}{3}$ A through each 12 Ω resistor. The current I_1 can also be found by applying the current divider rule as

$$I_1 = I_2 \left(\frac{\frac{1}{12}}{\frac{1}{12} + \frac{1}{12} + \frac{1}{12}} \right) = \frac{\frac{1}{12}}{\frac{1}{12} + \frac{1}{12} + \frac{1}{12}} = \frac{1}{3}\,\text{A} \qquad \blacksquare$$

CHAPTER 5

Linear Network Analysis

Our methods for solving circuit problems up to this point have included applying Ohm's law and Kirchhoff's laws, resistive circuit simplification, and the voltage and current divider rules. This approach works for all circuit problems, but as the circuit complexity increases, it becomes more difficult to solve problems. In this section, we introduce the node-voltage method and the mesh-current method to provide a systematic and easy solution of circuit problems. The application of the node-voltage method involves expressing the branch currents in terms of one or more node voltages, and applying KCL at each of the nodes. The application of the mesh-current method involves expressing the branch voltages in terms of mesh currents, and applying KVL around each mesh. These two methods are systematic approaches that lead to a solution that is efficient and robust, resulting in a minimum number of simultaneous equations that saves time and effort.

In both cases, the resulting linear set of simultaneous equations is solved to determine the unknown voltages or currents. The number of unknown voltages for the node-voltage method or currents for the mesh-current method determines the number of equations. The number of independent equations necessitates:

- N-1 equations involving KCL at N-1 nodes for the node-voltage method. This number may be fewer if there are voltage sources in the circuit.
- N-1 equations involving KVL around each of the meshes in the circuit for the mesh-current method. This number may be fewer if there are current sources in the circuit.

As we will see, MATLAB is ideal for solving problems that involves solution of simultaneous equations, providing a straightforward tool that minimizes the amount of work.

5.1 NODE-VOLTAGE METHOD

The use of node equations provides a systematic method for solving circuit analysis problems by the application of KCL at each essential node. The node-voltage method involves the following two steps:

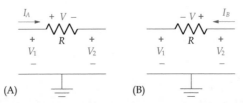

FIGURE 5.1: Ohm's law written in terms of node voltages.

1. Assign each node a voltage with respect to a reference node (ground). The reference node is usually the one with the most branches connected to it, and is denoted with the symbol ⏚. All voltages are written with respect to the reference node.

2. Except for the reference node, we write KCL at each of the N-1 nodes.

The current through a resistor is written using Ohm's law, with the voltage expressed as the difference between the potential on either end of the resistor with respect to the reference node as shown in Fig. 5.1. We express node-voltage equations as the currents leaving the node. Two adjacent nodes give rise to the current moving to the right (like Fig. 5.1A) for one node, and the current moving to the left (like Fig. 5.1B) for the other node. The current is written for (A) as $I_A = \frac{V}{R} = \frac{V_1 - V_2}{R}$ and for (B) as $I_B = \frac{V}{R} = \frac{V_2 - V_1}{R}$. It is easy to verify in (A) that $V = V_1 - V_2$ by applying KVL.

If one of the branches located between an essential node and the reference node contains an independent or dependent voltage source, we do not write a node equation for this node, because the node voltage is known. This reduces the number of independent node equations by one and the amount of work in solving for the node voltages. In writing the node equations for the other nodes, we write the value of the independent voltage source in those equations. Consider Fig. 5.1 (A) and assume the voltage V_2 results from an independent voltage source of 5 V. Since the node voltage is known, we do not write a node voltage equation for node 2 in this case. When writing the node-voltage equation for node 1, the current I_A is written as $I_A = \frac{V_1 - 5}{R}$. Ex. 5.1 further illustrates this case.

Example 5.1. Find V_1 using the node-voltage method.

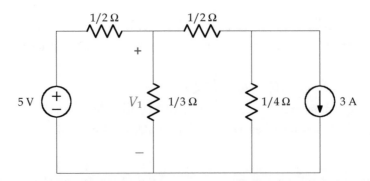

Solution. This circuit has two essential nodes, labeled 1 and 2 in the redrawn circuit that follows, with the reference node and two node voltages, V_1 and V_2, indicated. The node involving the 5 V voltage source has a known node voltage and therefore we do not write a node equation for it.

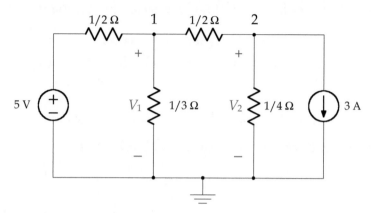

Summing the currents leaving node 1 gives

$$2(V_1 - 5) + 3V_1 + 2(V_1 - V_2) = 0$$

which simplifies to

$$7V_1 - 2V_2 = 10$$

Summing the currents leaving node 2 gives

$$2(V_2 - V_1) + 4V_2 + 3 = 0$$

which simplifies to

$$-2V_1 + 6V_2 = -3$$

The two node equations are written in matrix format as

$$\begin{bmatrix} 7 & -2 \\ -2 & 6 \end{bmatrix} \begin{bmatrix} V_1 \\ V_2 \end{bmatrix} = \begin{bmatrix} 10 \\ -3 \end{bmatrix}$$

and solved with MATLAB as follows:

```
≫ A = [7 − 2; −2 6];
≫ F = [10; −3];
≫ V = A\F
V =
     1.4211
    −0.0263
```

Thus, $V_1 = 1.4211$ V. ∎

Generally, the coefficient for the node voltage is the sum of the conductances connected to the node. The coefficients for the other node voltages are the negative of the sum of conductances connected between the node voltage and other node voltages. If the input consists of a set of current sources applied at each node, then the node equations have the following form.

$$
\begin{aligned}
G_{1,1}V_1 - G_{1,2}V_2 - \cdots - G_{1,N-1}V_{N-1} &= I_1 \\
-G_{2,1}V_1 + G_{2,2}V_2 - \cdots - G_{2,N-1}V_{N-1} &= I_2 \\
&\vdots \\
-G_{N-1,1}V_1 - G_{N-1,2}V_2 - \cdots - G_{N-1,N-1}V_{N-1} &= I_N
\end{aligned}
\tag{5.1}
$$

Equation (5.1) is put in matrix form for solution by MATLAB as

$$
\begin{bmatrix}
G_{1,1} & -G_{2,1} & \cdots & -G_{1,N-1} \\
-G_{2,1} & G_{2,2} & \cdots & -G_{2,N-1} \\
& \ddots & & \\
-G_{N-1,1} & -G_{N-1,2} & \cdots & G_{N-1,N-1}
\end{bmatrix}
\begin{bmatrix}
V_1 \\ V_2 \\ \vdots \\ V_{N-1}
\end{bmatrix}
=
\begin{bmatrix}
I_1 \\ I_2 \\ \vdots \\ I_{N-1}
\end{bmatrix}
\tag{5.2}
$$

Note the symmetry about the main diagonal where the off-diagonal terms are equal to each other and negative. This is true of all circuits without dependent sources. A dependent source destroys this symmetry. In general, if a circuit has dependent sources, the node-voltage approach is the same as before except for an additional equation describing the relationship between the dependent source and the node voltage. In cases involving more than one dependent source, there is one equation for each dependent source in terms of the node voltages.

Example 5.2.   For the following circuit, find V_3 using the node-voltage method.

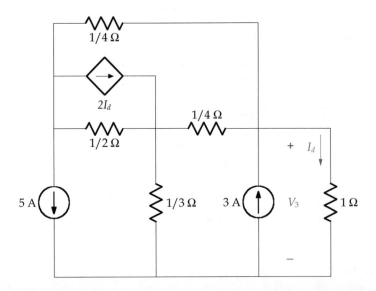

Solution. Notice that this circuit has three essential nodes and a dependent current source. We label the essential nodes 1, 2 and 3 in the redrawn circuit, with the reference node at the bottom of the circuit and three node voltages V_1, V_2 and V_3, as indicated.

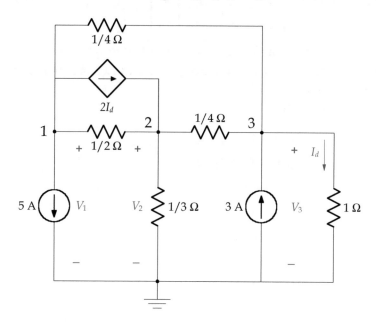

Note that $I_d = V_3$ according to Ohm's law. Summing the currents leaving node 1 gives

$$5 + 2(V_1 - V_2) + 2I_d + 4(V_1 - V_3) = 0$$

which reduces to

$$6V_1 - 2V_2 - 2V_3 = -5$$

Summing the currents leaving node 2 gives

$$-2I_d + 2(V_2 - V_1) + 3V_2 + 4(V_2 - V_3) = 0$$

which simplifies to

$$-2V_1 + 9V_2 - 6V_3 = 0$$

Summing the currents leaving node 3 gives

$$4(V_3 - V_2) - 3 + V_3 + 4(V_3 - V_1) = 0$$

reducing to

$$-4V_1 - 4V_2 + 9V_3 = 3$$

The three node equations are written in matrix format as

$$\begin{bmatrix} 6 & -2 & -2 \\ -2 & 9 & -6 \\ -4 & -4 & 9 \end{bmatrix} \begin{bmatrix} V_1 \\ V_2 \\ V_3 \end{bmatrix} = \begin{bmatrix} -5 \\ 0 \\ 3 \end{bmatrix}$$

Notice that the system matrix is no longer symmetrical because of the dependent current source, and two of the three nodes have a current source giving rise to a nonzero term on the right-hand side of the matrix equation.

Solving with MATLAB gives

$$\gg A = [6\,{-}2\,{-}2;\,{-}2\,9\,{-}\,6;\,{-}4\,{-}4\,9];$$
$$\gg F = [-5;\,0;\,3];$$
$$\gg V = A\backslash F$$
$$V =$$
$$-1.1471$$
$$-0.5294$$
$$-0.4118$$

Thus $V_3 = -0.4118\,\text{V}$. ∎

If one of the branches has an independent or controlled voltage source located between two essential nodes as shown in Fig. 5.2, the current through the source is not easily expressed in terms of node voltages. In this situation, we form a *supernode* by combining the two nodes. The supernode technique requires only one node equation in which the current, I_A, is passed through the source and written in terms of currents leaving node 2. Specifically, we replace I_A with $I_B + I_C + I_D$ in terms of node voltages. Because we have two unknowns and one supernode equation, we write a second equation by applying KVL for the two node voltages 1

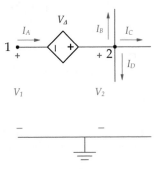

FIGURE 5.2: A dependent voltage source is located between nodes 1 and 2.

and 2 and the source as

$$-V_1 - V_\Delta + V_2 = 0$$

or

$$V_\Delta = V_1 - V_2$$

Example 5.3. For the following circuit, find V_3.

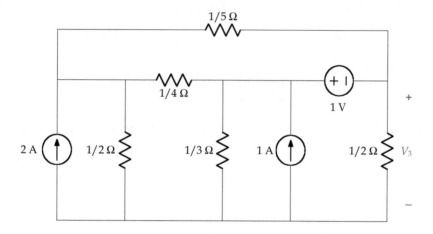

Solution. The circuit has three essential nodes, two of which are connected to an independent voltage source and form a supernode. We label the essential nodes as 1, 2 and 3 in the redrawn circuit, with the reference node at the bottom of the circuit and three node voltages, V_1, V_2 and V_3 as indicated.

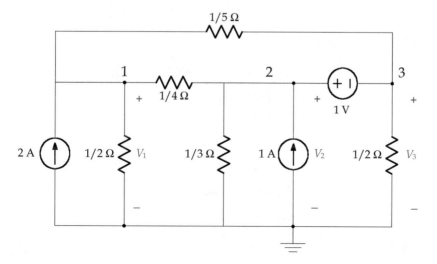

Summing the currents leaving node 1 gives

$$-2 + 2V_1 + 5(V_1 - V_3) + 4(V_1 - V_2) = 0$$

Simplifying gives

$$11V_1 - 4V_2 - 5V_3 = 2$$

Nodes 2 and 3 are connected by an independent voltage source, so we form a supernode $2 + 3$. Summing the currents leaving the supernode $2 + 3$ gives

$$4(V_2 - V_1) + 3V_2 - 1 + 2V_3 + 5(V_3 - V_1) = 0$$

Simplifying yields

$$-9V_1 + 7V_2 + 7V_3 = 1$$

The second supernode equation is KVL through the node voltages and the independent source, giving

$$-V_2 + 1 + V_3 = 0$$

or

$$-V_2 + V_3 = -1$$

The two node and KVL equations are written in matrix format as

$$\begin{bmatrix} 11 & -4 & -5 \\ -9 & 7 & 7 \\ 0 & -1 & 1 \end{bmatrix} \begin{bmatrix} V_1 \\ V_2 \\ V_3 \end{bmatrix} = \begin{bmatrix} 2 \\ 1 \\ -1 \end{bmatrix}$$

Solving with MATLAB gives

```
>> A = [11 −4 −5; −9 7 7; 0 −1 1];
>> F = [2; 1; −1];
>> V = A\F
V =
        0.4110
        0.8356
       −0.1644
```

Thus $V_3 = -0.1644$.   ■

5.2 MESH-CURRENT METHOD

Another method for analyzing planar circuits is called the mesh-current method. A mesh is a closed path without any other closed paths within it and a planar circuit is a circuit without any overlapping branches. All of our problems in this book involve planar circuits. The mesh-current method provides a systematic process for solving circuit analysis problems with the application of KVL around each mesh. The mesh-current method involves the following two steps:

1. Define the mesh currents in the circuit. By convention, we draw the mesh currents with a circle, arc or a surface on the inside perimeter of the mesh. Moreover, we define the mesh-current direction for all meshes to be clockwise.

2. Write a set of mesh equations using KVL. In general, we write one equation for each mesh. Under special circumstances, the number of mesh equations may be fewer than the total number of meshes.

In writing the mesh equation, we move through the mesh in a clockwise direction by writing the voltage drops in terms of mesh currents. Whenever a circuit element is shared by two meshes, as in Fig. 5.3, the voltage drop across the resistor is

$$V = RI = R(I_1 - I_2)$$

when writing the equation for mesh 1. When writing the mesh equation for mesh 2, the clockwise direction gives a voltage drop according to convention as $R(I_2 - I_1)$, just the opposite as in mesh 1. Moreover, according to KCL

$$I = I_1 - I_2$$

Mesh currents, as in Fig. 5.3, are not measurable with an ammeter in that they do not equal a branch current. The current through the branch is composed of the difference between the two mesh currents, here $I = I_1 - I_2$. Even though the mesh current is not real, it is a powerful technique that simplifies analysis of circuit problems as shown in the next example.

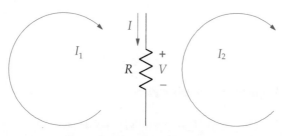

FIGURE 5.3: Mesh currents.

Example 5.4.   Find I in the following circuit.

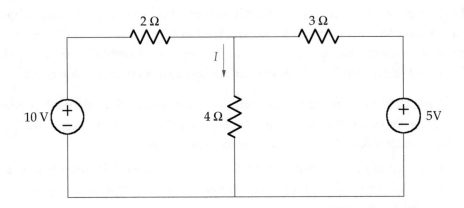

Solution.   There are two meshes in this circuit. Mesh currents are always defined in a clockwise direction, one for each mesh, as illustrated in the redrawn circuit that follows.

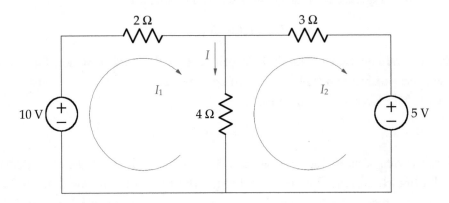

Summing the voltage drops around mesh 1 gives

$$-10 + 2I_1 + 4(I_1 - I_2) = 0$$

which simplifies to

$$6I_1 - 4I_2 = 10$$

Summing the voltage drops around mesh 2 gives

$$4(I_2 - I_1) + 3I_2 + 5 = 0$$

which simplifies to

$$-4I_1 + 7I_2 = -5$$

The two mesh equations are written in matrix form as

$$\begin{bmatrix} 6 & -4 \\ -4 & 7 \end{bmatrix} \begin{bmatrix} I_1 \\ I_2 \end{bmatrix} = \begin{bmatrix} 10 \\ -5 \end{bmatrix}$$

Comments made in Section 5.1 on the structure and symmetry of the system matrix with the node-voltage method also apply for the mesh-current method. We solve this problem with MATLAB, giving

$$\gg A = [6 -4; -4\ 7];$$
$$\gg F = [10; -5];$$
$$\gg I = A\backslash F$$
$$I =$$
$$1.9231$$
$$0.3846$$

The current I is found from $I = I_1 - I_2 = 1.9231 - 0.3846 = 1.5385$ A.   ∎

When one of the branches has an independent or dependent current source in the circuit, a modification must be made to the mesh-current method. Depending on whether the current source is on the outer perimeter or inside the circuit, as shown in Fig. 5.4, we handle both cases as follows:

1. The current source is located on the perimeter, as in the circuit on the left in Fig. 5.4, where the mesh current equals the branch current. In this case, we do not write a mesh equation because the current is known. The mesh current I_1 equals the source current,

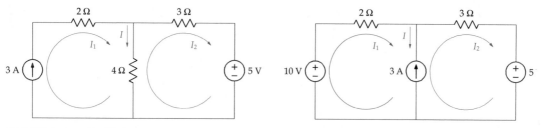

FIGURE 5.4: (Left) A current source on the perimeter of a circuit. (Right) A current source in a branch between two meshes.

$I_1 = 3$ A. The equation for mesh 2 is found by applying KVL, giving

$$4(I_2 - I_1) + 3I_2 + 5 = 4(I_2 - 3) + 3I_2 + 5 = 0,$$

which gives $I_2 = 1$ A.

2. The current source is located within the circuit where the mesh current does not equal the branch current as shown in Fig. 5.4 (right). Since we cannot easily write the voltage drop across the current source, we form a *supermesh*. A supermesh is formed by combing two meshes together, with one equation describing both meshes. In this case, the equation starts wit the first mesh and continues on to the second mesh, circumventing the voltage drop across the current source. Here the supermesh equation is

$$-10 + 2I_1 + 3I_2 + 5 = 0$$

Because there are two unknown currents, we need two independent equations. The second equation is written using KCL for the current source and the two mesh currents. Here the KCL equation is $I_2 - I_1 = 3$.

Example 5.5. Find V_0 as shown in the following circuit.

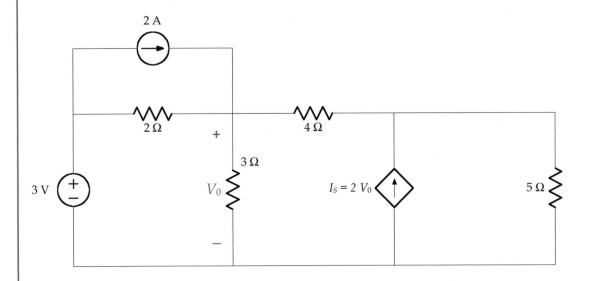

Solution. This circuit has four meshes, as shown in the circuit diagram that follows. Mesh 4 has a current source on the perimeter, so we do not write a mesh equation for it, but write simply $I_4 = 2$ A.

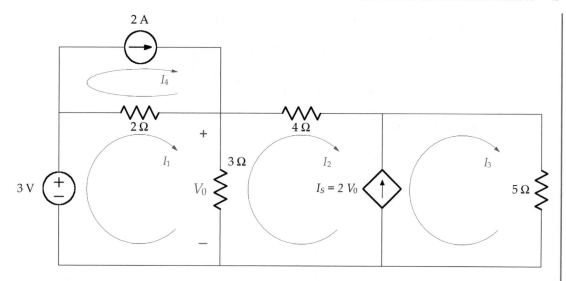

Summing the voltages around mesh 1 gives

$$-3 + 2\left(I_1 - 2\right) + 3(I_1 - I_2) = 0$$

Simplifying gives

$$5I_1 - 3I_2 = 7$$

Since there is a dependent current source inside the circuit, we form a supermesh for meshes 2 and 3. Summing the voltages around supermesh $2 + 3$ yields

$$3\left(I_2 - I_1\right) + 4I_2 + 5I_3 = 0$$

which reduces to

$$-3I_1 + 7I_2 + 5I_3 = 0$$

Applying KCL for the dependent current source gives

$$2V_0 = 2 \times 3\left(I_1 - I_2\right) = I_3 - I_2$$

yielding

$$6I_1 - 5I_2 - I_3 = 0$$

The three independent equations are written in matrix format as

$$\begin{bmatrix} 5 & -3 & 0 \\ -3 & 7 & 5 \\ 6 & -5 & -1 \end{bmatrix} \begin{bmatrix} I_1 \\ I_2 \\ I_3 \end{bmatrix} = \begin{bmatrix} 7 \\ 0 \\ 0 \end{bmatrix}$$

Solution using MATLAB gives

```
≫ A = [5 − 3 0; −3 7 5; 6 − 5 − 1];
≫ F = [7; 0; 0];
≫ I = A\F
I =
      14.0000
      21.0000
     −21.0000
```

Thus,

$$V_0 = 3(I_1 - I_2) = 3(14 - 21) = 21 \text{ V} \qquad \blacksquare$$

5.3 LINEARITY, SUPERPOSITION AND SOURCE TRANSFORMATIONS

5.3.1 Linearity and Superposition

If a linear system is excited by two or more independent sources, then the total response is the sum of the separate individual responses to each input. This property is called the principle of *superposition*. Specifically for circuits, the response to several independent sources is the sum of responses to each independent source with the other independent sources dead, where

- A dead voltage source is a short circuit
- A dead current source is an open circuit

In linear circuits with multiple independent sources, the total response is the sum of each independent source taken one at a time. This analysis is carried out by removing all of the sources except one, and assuming the other sources are dead. After the circuit is analyzed with the first source, it is set equal to a dead source and the next source is applied with the remaining sources dead. When each of the sources have been analyzed, the total response is obtained by summing the individual responses. Note carefully that this principle holds true solely for independent sources. Dependent sources must remain in the circuit when applying this technique, and they must be analyzed based on the current or voltage for which it is defined. It should be apparent that voltages and currents in one circuit differ among circuits, and that we cannot mix and match voltages and currents from one circuit with another.

Generally, superposition provides a simpler solution than is obtained by evaluating the total response with all of the applied sources. This property is especially valuable when dealing with an input consisting of a pulse or delays. These are considered in future sections.

Example 5.6. Using superposition, find V_0 as shown in the following figure.

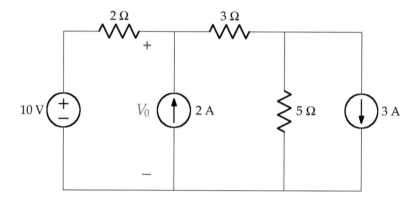

Solution. We start by analyzing the circuit with just the 10 V source active and the two current sources dead, as shown in the following figure.

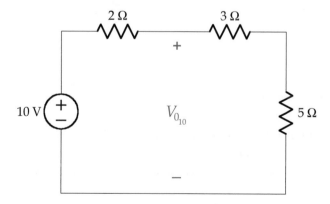

The voltage divider rule easily gives the response, $V_{0_{10}}$, due to the 10 V source

$$V_{0_{10}} = 10 \left(\frac{8}{2 + 8} \right) = 8 \text{ V}$$

Next consider the 2 A source active, and the other two sources dead, as shown in the following circuit.

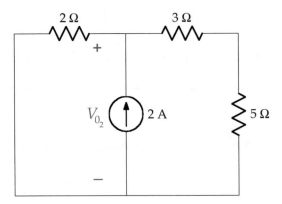

Combining the resistors in an equivalent resistance, $R_{EQ} = 2 \parallel (3 + 5) = \frac{2 \times 8}{2+8} = 1.6\,\Omega$, and then applying Ohm's law gives $V_{0_2} = 2 \times 1.6 = 3.2\,\text{V}$.

Finally, consider the response, V_{0_3}, to the 3 A source as shown in the following figure.

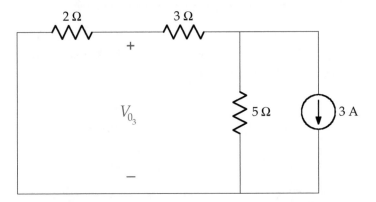

To find V_{0_3}, note that the 3 A current splits into 1.5 A through each branch ($2 + 3\,\Omega$ and $5\,\Omega$), and $V_{0_3} = -1.5 \times 2 = -3\,\text{V}$.

The total response is given by the sum of the individual responses as

$$V_0 = V_{0_{10}} + V_{0_2} + V_{0_3} = 8 + 3.2 - 3 = 8.2\,\text{V}$$

This is the same result we would have found if we analyzed the original circuit directly using the node-voltage or mesh-current method.   ∎

Example 5.7. Find the voltage across the 5 A current source, V_5, in the following figure using superposition.

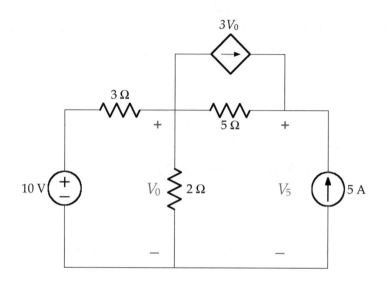

Solution. First consider finding the response, $V_{0_{10}}$, due to the 10 V source only with the 5 A source dead as shown in the following figure. As required during the analysis, the dependent current source is kept in the modified circuit and should not be set dead.

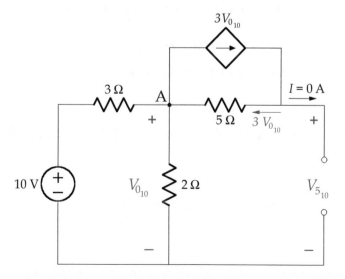

Notice that no current flows through the open circuit created by the dead current source, and that the current flowing through the 5 Ω resistor is $3V_0$. Therefore, applying KCL at node A gives

$$\frac{V_{0_{10}} - 10}{3} + \frac{V_{0_{10}}}{2} + 3V_{0_{10}} - 3V_{0_{10}} = 0$$

which gives $V_{0_{10}} = 4$ V. KVL gives $-V_{0_{10}} - 5 \times 3V_{0_{10}} + V_{5_{10}} = 0$, and therefore $V_{5_{10}} = 64$ V.
Next consider finding the response, V_{0_5}, due to the 5 A source, with the 10 V source dead.

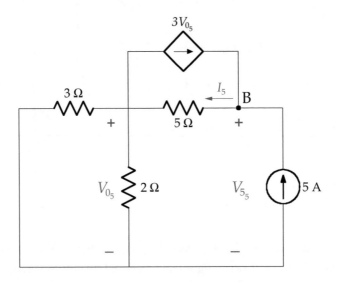

First combine the two resistors in parallel (3 Ω || 2 Ω), giving 1.2 Ω. V_{0_5} is easily calculated by Ohm's law as $V_{0_5} = 5 \times 1.2 = 6$ V. KCL is then applied at node B to find I_5, giving

$$-3V_{0_5} + I_5 - 5 = 0$$

With $V_{0_5} = 6$ V, $I_5 = 3 \times 6 + 5 = 23$ A. Finally, apply KVL around the closed path

$$-V_{0_5} - 5I_5 + V_{5_5} = 0$$

or $V_{5_5} = V_{0_5} + 5I_5 = 6 + 5 \times 23 = 121$ V. The total response is given by the sum of the individual responses as

$$V_5 = V_{5_{10}} + V_{5_5} = 64 + 121 = 185 \text{ V} \qquad \blacksquare$$

5.3.2   Equivalent Sources

We call two sources equivalent if they each produce the same voltage and current regardless of the resistance. Consider the two circuits in Fig. 5.5. If $I_s = \frac{V_s}{R_s}$ as shown in the figure on the right, then the same current and voltage are seen in resistor R_l in either circuit as easily shown

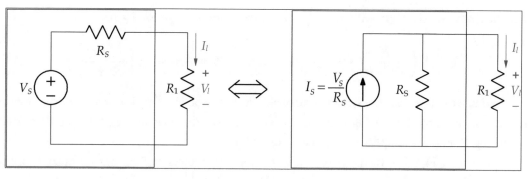

FIGURE 5.5: Two equivalent circuits.

using voltage and current divider rules. For the circuit on the left, the current and voltage for R_l are

$$V_l = V_s \left(\frac{R_l}{R_l + R_s} \right) \quad \text{and} \quad I_l = \frac{V_s}{R_l + R_s}$$

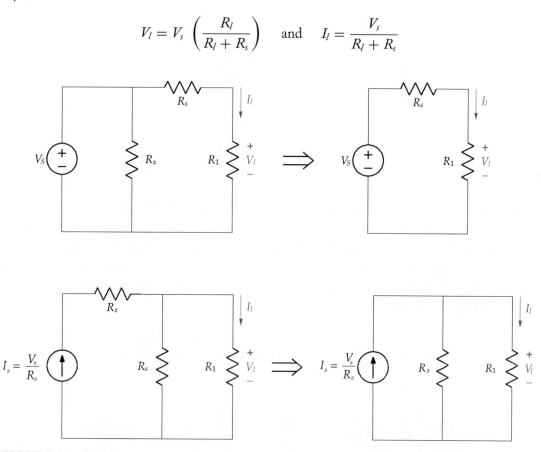

FIGURE 5.6: Equivalent circuits. In both circuits on the left, the resistor R_x has no effect on the circuit and can be removed.

For the circuit on the right with $I_s = \frac{V_s}{R_s}$, the current and voltage for R_l are

$$I_l = I_s \left(\frac{R_s}{R_l + R_s} \right) = \frac{V_s}{R_l + R_s} \quad \text{and} \quad V_l = I_l R_l = V_s \left(\frac{R_l}{R_l + R_s} \right)$$

Therefore, we can replace the voltage source and R_s in the box in Fig. 5.5 (left) with the current source and R_s in the box in Fig. 5.5 (right). We shall see that exchanging source plus resistor according to Fig. 5.5 simplifies the analysis of circuits.

Consider Fig. 5.6. In the two circuits on the left, the resistor R_x has no impact on the voltage and current on R_l, and as indicated, these circuits can be replaced by the two circuits on the right by completely removing R_x.

Example 5.8. Use source transformations to find V_0 in the following figure.

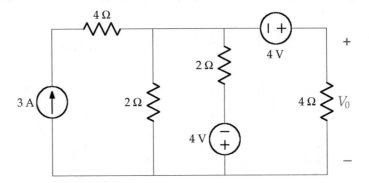

Solution. Our strategy in this solution involves combining resistors in series and parallel, current sources in parallel and voltage sources in series. We first remove the $4\,\Omega$ resistor since it is in series with the 3 A source and has no effect on the circuit, and transform the $2\,\Omega$ resistor and 4 V source into a $\frac{4}{2} = 2$ A source in parallel with a $2\,\Omega$ resistor as shown in the following figure. Notice that the current direction in the transformed source is in agreement with the polarity of the 4 V source.

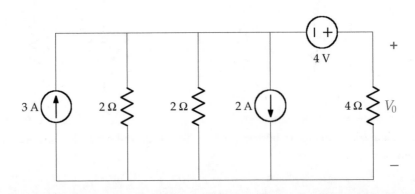

As shown in the next figure, combining the two parallel current sources results in a 1 A source, and combining the two parallel resistors results in a 1 Ω resistance.

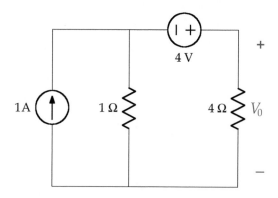

Another source transformation is carried out on the current source and resistor in parallel as shown in the next figure.

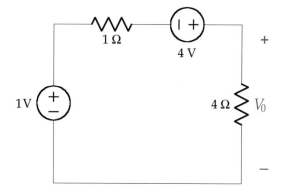

The two voltage sources are combined, resulting in a 5 V source. Using the voltage divider gives

$$V_0 = 5 \left(\frac{4}{4+1} \right) = 4\,\text{V} \qquad \blacksquare$$

CHAPTER 6

Thévenin's and Norton's Theorems

Any combination of resistances, controlled sources, and independent sources with two external terminals (A and B, denoted A,B) can be replaced by a single resistance and an independent source, as shown in Fig. 6.1. A Thévenin equivalent circuit reduces the original circuit into a voltage source in series with a resistor (upper right of Fig. 6.1). A Norton equivalent circuit reduces the original circuit into a current source in parallel with a resistor (lower right of Fig. 6.1). These two theorems help reduce complex circuits into simpler circuits. We refer to the circuit elements connected across the terminals A,B (that are not shown) as the *load*. The Thévenin equivalent circuit and Norton equivalent circuits are equivalent to the original circuit in that the same voltage and current are observed across any load. Usually, the load is not included in the simplification because it is important for other analysis, such as maximum power expended by the load. Although we focus here on sources and resistors, these two theorems can be extended to any circuit composed of linear elements with two terminals.

6.1 THÉVENIN'S THEOREM

Thévenin's Theorem states that an equivalent circuit consisting of an ideal voltage source, V_{OC}, in series with an equivalent resistance, R_{EQ}, can be used to replace any circuit that consists of independent and dependent voltage and current sources and resistors. V_{OC} is equal to the open circuit voltage across terminals A,B as shown in Fig. 6.2, and calculated using standard techniques such as the node-voltage or mesh-current methods.

The resistor R_{EQ} is the resistance seen across the terminals A,B when all sources are *dead*. Recall that a dead voltage source is a short circuit and a dead current source is an open circuit.

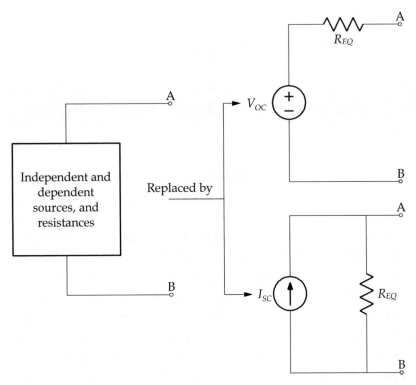

FIGURE 6.1: A general circuit consisting of independent and dependent sources can be replaced by a voltage source (V_{OC}) in series with a resistor (R_{EQ}) or a current source (I_{SC}) in parallel with a resistor (R_{EQ}).

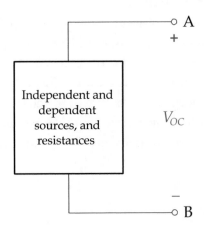

FIGURE 6.2: The open circuit voltage, V_{OC}, is calculated across the terminals A,B using standard techniques such as node-voltage or mesh-current methods.

Example 6.1   Find the Thévenin equivalent circuit with respect to terminals A,B for the following circuit.

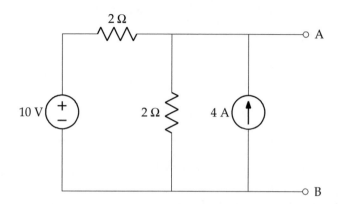

Solution.   The solution to finding the Thévenin equivalent circuit is done in two parts, first finding V_{OC} and then solving for R_{EQ}. The open circuit voltage, V_{OC}, is easily found using the node-voltage method as shown in the following circuit.

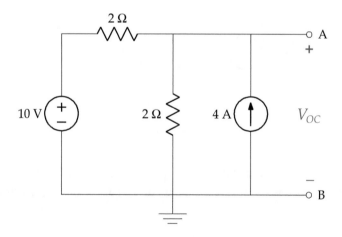

The sum of currents leaving the node is

$$\frac{V_{OC} - 10}{2} + \frac{V_{OC}}{2} - 4 = 0$$

and V_{OC} = 9 V.

Next, R_{EQ} is found by first setting all sources dead (the current source is an open circuit and the voltage source is a short circuit), and then finding the resistance seen from the terminals A,B as shown in the following figure.

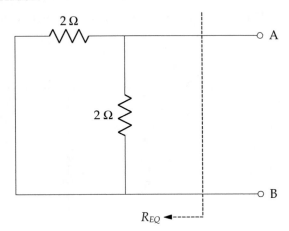

From the previous circuit, it is clear that R_{EQ} is equal to $1\,\Omega$ (that is, $2\,\Omega \parallel 2\,\Omega$). Thus, the Thévenin equivalent circuit is

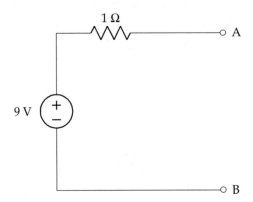

It is important to note that the circuit used in finding V_{OC} is not to be used in finding R_{EQ} as not all voltages and currents are relevant in the other circuit and one cannot simply mix and match. ∎

6.2 NORTON'S THEOREM

Norton's Theorem states that an equivalent circuit consisting of an ideal current source, I_{SC}, in parallel with an equivalent resistance, R_{EQ}, can be used to replace any circuit that consists of independent and dependent voltage and current sources and resistors. I_{SC} is equal to the current flowing through a short between terminals A,B as shown in Fig. 6.3, and calculated using standard techniques such as the node-voltage or mesh-current methods. The Thévenin

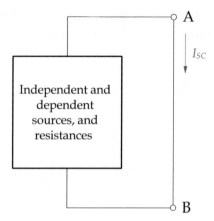

FIGURE 6.3: The short circuit current, I_{SC}, is calculated by placing a short across the terminals A,B, and finding the current through the short using standard techniques such as node-voltage or mesh-current methods.

and Norton equivalent circuits are related to each other according to

$$R_{EQ} = \frac{V_{OC}}{I_{SC}} \qquad (6.1)$$

This is easily seen by shorting the terminals A,B in Fig. 6.2. The current flowing through the short is $\frac{V_{OC}}{I_{SC}}$ in agreement with Eq. (6.1). It should be clear that Figs. 6.2 and 6.3 are source transformations of each other.

Example 6.2. Find the Norton equivalent circuit with respect to terminals A,B for the following circuit.

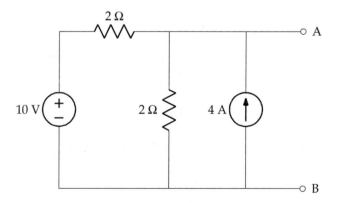

Solution. This circuit is the same as Ex. 6.1, so $R_{EQ} = 1\,\Omega$. To find I_{SC} we place a short between the terminals A,B as shown in the following figure.

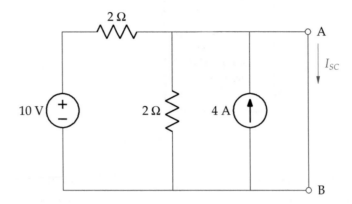

Note that the $2\,\Omega$ resistor is in parallel with the short across terminals A,B, therefore it is removed since no current flows through it, as shown in the next figure.

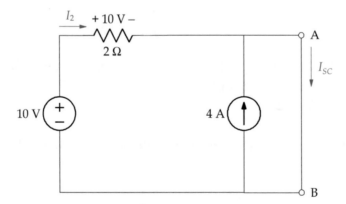

Note also that current I_2 equals 5 A since the 10 V source is applied directly across the $2\,\Omega$ resistor. Applying KCL at the node denoted A, gives

$$0 = -I_2 - 4 + I_{SC} = -5 - 4 + I_{SC}$$

and $I_{SC} = 9\,\text{A}$. We could have found the current I_{SC} directly from the solution of Ex. 6.1 since

$$I_{SC} = \frac{V_{OC}}{R_{EQ}} = \frac{9}{1} = 9\,\text{A}$$

The Norton equivalent circuit is shown in the following figure.

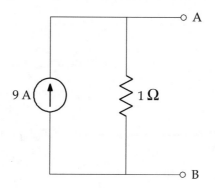

6.3   DEPENDENT SOURCES AND THÉVENIN AND NORTON EQUIVALENT CIRCUITS

If a circuit connected to terminals A,B contains dependent sources, the process for finding R_{EQ} for either the Thévenin or Norton equivalent circuit must be modified, because the dependent source has resistance, and this resistance cannot be calculated when setting all sources dead.

In this case we find V_{OC} and I_{SC}, and then calculate $R_{EQ} = \frac{V_{OC}}{I_{SC}}$.

Example 6.3 Find the Thévenin equivalent circuit with respect to terminals A,B for the following circuit.

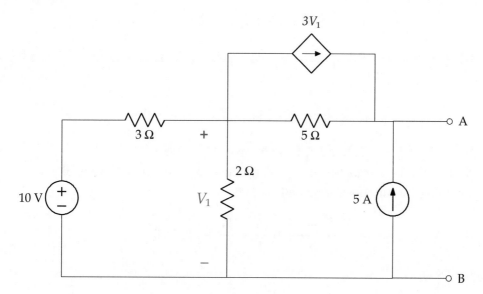

Solution. Since this circuit has a dependent source, R_{EQ} is found by $R_{EQ} = \frac{V_{OC}}{I_{SC}}$. Let's first find V_{OC} using the mesh-current method as labeled in the next figure.

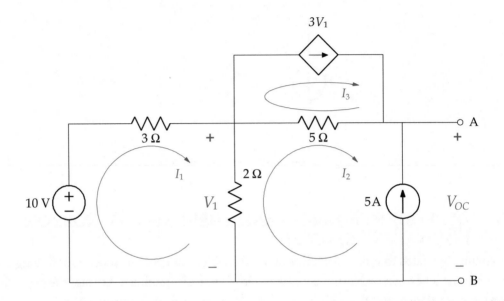

Note that since there are two current sources on the perimeter of the circuit, $I_2 = -5\,\text{A}$ and $I_3 = 3V_1 = 3 \times 2\,(I_1 - I_2) = 6I_1 + 30$.

Summing the voltage drops around mesh 1 gives

$$-10 + 3I_1 + 2\,(I_1 - I_2) = 0$$

Simplifying with $I_2 = -5\,\text{A}$ gives $I_1 = 0\,\text{A}$. Applying KVL around mesh 2 to find V_{OC} gives

$$2\,(I_2 - I_1) + 5\,(I_2 - I_3) + V_{OC} = 0$$

With $I_1 = 0\,\text{A}, \quad I_2 = -5\,\text{A}$ and $I_3 = 30\,\text{A}$, gives $V_{OC} = 185\,\text{V}$.

Next, we find I_{SC} using the node-voltage method as labeled in the next figure.

Summing the currents leaving node Δ gives

$$\frac{V_1 - 10}{3} + \frac{V_1}{2} + \frac{V_1}{5} + 3V_1 = 0$$

and $V_1 = \frac{100}{121}\,\text{V}$.

Applying KCL at terminal A gives

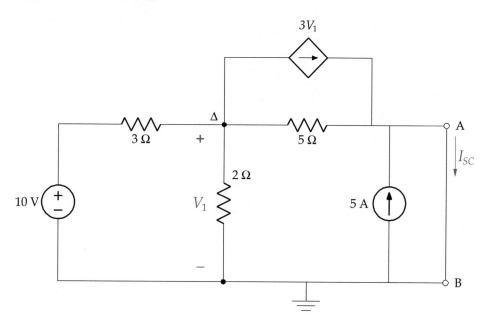

$$-3V_1 - 5 - \frac{V_1}{5} + I_{SC} = 0$$

With $V_1 = \frac{100}{121}$, gives $I_{SC} = 7.65$ A. Therefore $R_{EQ} = \frac{V_{OC}}{I_{SC}} = \frac{185}{7.65} = 24.18\ \Omega$. The Thévenin equivalent circuit is

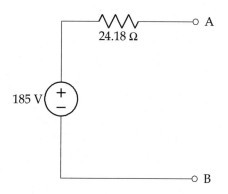

CHAPTER 7

Inductors

In the previous sections, we considered circuits involving sources and resistors that are described with algebraic equations. Any changes in the source are instantaneously observed in the response. In this section, we examine the inductor, a passive element that relates the voltage–current relationship with a differential equation. Circuits that contain inductors are written in terms of derivatives and integrals. Any changes in the source with circuits that contain inductors, i.e., a step input, have a response that is not instantaneous, but has natural response that changes exponentially and a forced response that is the same form as the source.

An inductor is a passive element that is able to store energy in a magnetic field, and is made by winding a coil of wire around a core that is an insulator or a ferromagnetic material. A magnetic field is established when current flows through the coil. We use the symbol ⌒⌒⌒⌒ to represent the inductor in a circuit; the unit of measure for inductance is the henry or henries (H), where 1 H = 1 V s/A. The relationship between voltage and current for an inductor is given by

$$v = L\frac{di}{dt} \tag{7.1}$$

The convention for writing the voltage drop across an inductor is similar to that of a resistor, as shown in Fig. 7.1.

Physically, current cannot change instantaneously through an inductor since an infinite voltage is required according to Eq. (7.1) (i.e., the derivative of current at the time of the instantaneous change is infinity). Mathematically, a step change in current through an inductor is possible by applying a voltage that is a Dirac delta function. For convenience, when a circuit has just DC currents (or voltages), the inductors can be replaced by short circuits since the voltage drop across the inductors are zero.

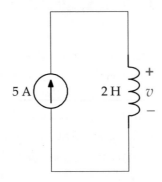

FIGURE 7.1: An inductor.

Example 7.1.   Find v in the following circuit.

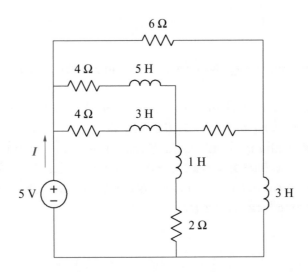

Solution.   Accordingly,

$$v = L\frac{di}{dt} = 2 \times \frac{d\,(5)}{dt} = 0$$

This example shows that an inductor acts like a short circuit to DC current.   ■

Example 7.2.   Find I in the following circuit, given that the source has been applied for a very long time.

Solution. Since the source has been applied for a very long time, only DC current flows in the circuit and the inductors can be replaced by short circuits as shown in the following figure.

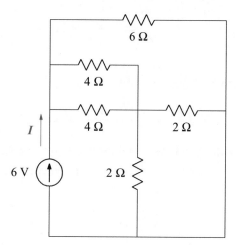

To find I, we find the total resistance seen by the source and use Ohm's law as follows:

$$R_{EQ} = 6 \parallel ((4 \parallel 4) + (2 \parallel 2)) = 2\,\Omega$$

therefore,

$$I = \frac{6}{2} = 3\,\Omega \qquad \blacksquare$$

Example 7.3. Find v in the following circuit.

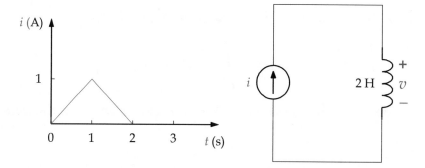

Solution. The solution to this problem is best approached by breaking it up into time intervals consistent with the changes in input current. Clearly for $t < 0$ and $t > 2$, the current is zero and therefore $v = 0$. We use Eq. (7.1) to determine the voltage in the other two intervals as follows.

For $0 < t < 1$

In this interval, the input is $i = t$, and

$$v = L\frac{di}{dt} = 2\frac{d(t)}{dt} = 2\,\text{V}$$

For $1 \leq t \leq 2$

In this interval, the input is $i = -(t - 2)$, and

$$v = L\frac{di}{dt} = 2\frac{d(-(t-2))}{dt} = -2\,\text{V} \qquad\qquad \blacksquare$$

Equation (7.1) defines the voltage across an inductor for a given current. Suppose one is given a voltage across an inductor and asked to find the current. We start from Eq. (7.1) by multiplying both sides by dt, giving

$$v(t)\,dt = L\,di$$

Integrating both sides yields

$$\int_{t_0}^{t} v(\lambda)\,d\lambda = L\int_{i(t_0)}^{i(t)} d\alpha$$

or

$$i(t) = \frac{1}{L}\int_{t_0}^{t} v(\lambda)\,d\lambda + i(t_0) \qquad\qquad (7.2)$$

For $t_0 = 0$, as is often the case in solving circuit problems, Eq. (7.2) reduces to

$$i(t) = \frac{1}{L}\int_{0}^{t} v(\lambda)\,d\lambda + i(0) \qquad\qquad (7.3)$$

and for $t_0 = -\infty$, the initial current is by definition equal to zero, and therefore Eq. (7.2) reduces to

$$i(t) = \frac{1}{L}\int_{-\infty}^{t} v(\lambda)\,d\lambda \qquad\qquad (7.4)$$

The initial current in Eq. (7.2), $i(t_0)$, is usually defined in the same direction as i, which means $i(t_0)$ is a positive quantity. If the direction of $i(t_0)$ is in the opposite direction of i (as will happen when we write node equations), then $i(t_0)$ is negative.

Example 7.4. Find i for $t \geq 0$ if $i(0) = 2$ A and $v(t) = 4e^{-3t}u(t)$ in the following circuit.

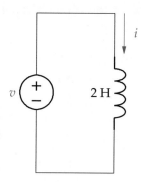

Solution. From Eq. (7.2), we have

$$i(t) = \frac{1}{L} \int_{t_0}^{t} v d\lambda + i(t_0) = \frac{1}{2} \int_{0}^{t} 4e^{-3\lambda} d\lambda + i(0) = \frac{1}{2} \int_{0}^{t} 4e^{-3\lambda} d\lambda + 2$$

$$= 2 \left. \frac{e^{-3\lambda}}{-3} \right|_{\lambda=0}^{t} + 2$$

$$= \frac{2}{3} \left(4 - e^{-3t} \right) u(t) \text{V} \qquad \blacksquare$$

7.1 POWER AND ENERGY

Since the inductor is a passive element, it absorbs power according to the relationship

$$p = vi = Li \frac{di}{dt} \qquad (7.5)$$

Within the inductor, power is not consumed as heat, as happens in a resistor, but stored as energy in the magnetic field around the coil during any period of time $[t_0, t]$ as

$$w(t) - w(t_0) = \int_{t_0}^{t} p \, dt = L \int_{t_0}^{t} i \frac{di}{dt} dt \qquad (7.6)$$

$$= L \int_{i(t_0)}^{i(t)} i \, di = \frac{1}{2} L \left([i(t)]^2 - [i(t_0)]^2 \right)$$

where the unit of energy is joules (J). At time equal to negative infinity, we assume that the initial current is zero and thus the total energy is given by

$$w(t) = \frac{1}{2} L i^2 \qquad (7.7)$$

Any energy stored in the magnetic field is recoverable. If power is negative in Eq. (7.5), energy is being extracted from the inductor. If power is positive, energy is being stored in the inductor.

Example 7.5.   Find the inductor power in the following circuit for $t > 0$ when $i(0) = 0$ A.

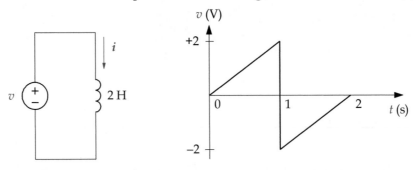

Solution.   We first solve for current so that Eq. (7.5) can be applied to find the power. As before, the solution is best approached by breaking it up into time intervals consistent with the changes in input voltage.

For $0 < t < 1$
In this interval, the input is $v = 2t$, and

$$i(t) = \frac{1}{L} \int_0^t v(\lambda)\, d\lambda + i(0) = \frac{1}{2} \int_0^t 2\lambda\, d\lambda = \frac{t^2}{2}\ \text{A}$$

Power for this interval is

$$p = vi = 2t \times \frac{t^2}{2} = t^3\ \text{W}$$

The current at $t = 1$ needed for the initial condition in the next part is

$$i(1) = \left.\frac{t^2}{2}\right|_{t=1} = \frac{1}{2}\ \text{A}$$

For $1 \le t \le 2$
In this interval, the input is $v = 2(t - 2)$, and

$$i(t) = \frac{1}{L} \int_1^t v(\lambda)d\lambda + i(1) = \frac{1}{2} \int_1^t 2(\lambda - 2)d\lambda + \frac{1}{2}$$

$$= \frac{t^2}{2} - 2t + 2\ \text{A}$$

Accordingly, power is

$$p = vi = 2(t-2) \times \left(\frac{t^2}{2} - 2t + 2 \right) = t^3 - 6t^2 + 12t - 8 \text{ W}$$

For $t > 2$

Power is zero in this interval since the voltage is zero. ∎

CHAPTER 8

Capacitors

A capacitor is a device that stores energy in an electric field by charge separation when appropriately polarized by a voltage. Simple capacitors consist of parallel plates of conducting material that are separated by a gap filled with a dielectric material. Dielectric materials, that is, air, mica, or Teflon, contain a large number of electric dipoles that become polarized in the presence of an electric field. The charge separation caused by the polarization of the dielectric is proportional to the external voltage and given by

$$q(t) = C\,v(t) \qquad (8.1)$$

where C represents the capacitance of the element. The unit of measure for capacitance is the farad or farads (F), where 1 F = 1 C/V. We use the symbol $\overset{\perp}{\top}C$ to denote a capacitor; most capacitors are measured in terms of microfarads (1 μF = 10^{-6} F) or picofarads (1 pF = 10^{-12} F). Figure 8.1 illustrates a capacitor in a circuit.

Using the relationship between current and charge, Eq. (8.1) is written in a more useful form for circuit analysis problems as

$$i = \frac{dq}{dt} = C\frac{dv}{dt} \qquad (8.2)$$

The capacitance of a capacitor is determined by the permittivity of the dielectric ($\varepsilon = 8.854 \times 10^{-12}$ F/M for air) that fills the gap between the parallel plates, the size of the gap between the plates, d, and the cross-sectional area of the plates, A, as

$$C = \frac{\varepsilon A}{d} \qquad (8.3)$$

As described, the capacitor physically consists of two conducting surfaces that stores charge, separated by a thin insulating material that has a very large resistance. In actuality, current does not flow through the capacitor plates. Rather, as James Clerk Maxwell hypothesized when he described the unified electromagnetic theory, a displacement current flows internally between capacitor plates and this current equals the current flowing into the capacitor and out of the capacitor. Thus KCL is maintained. It should be clear from Eq. (8.2) that dielectric

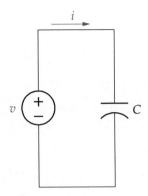

FIGURE 8.1: Circuit with a capacitor.

materials do not conduct DC currents; capacitors act as open circuits when DC currents are present.

Example 8.1. Suppose $v = 5$ V and $C = 2$ F for the circuit shown in Fig. 8.1 Find i.

Solution.

$$i = C\frac{dv}{dt} = 2 \times \frac{d}{dt}(5) = 0$$

A capacitor is an open circuit to DC voltage. ∎

Example 8.2. Find I in the following circuit given that the current source has been applied for a very long time.

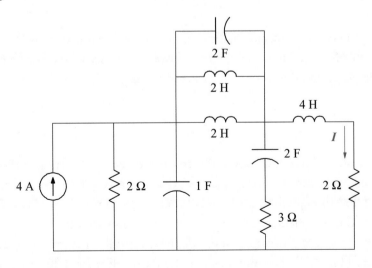

Solution. Since the source has been applied for a very long time, only DC current flows in the circuit. Furthermore the inductors can be replaced by short circuits and capacitors can be replaced by open circuits, as shown in the following figure.

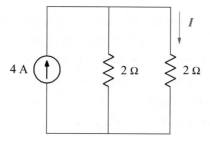

To find I, we use the current divider law as

$$I = 4 \times \frac{\frac{1}{2}}{\frac{1}{2} + \frac{1}{2}} = 2 \text{ A}$$ ∎

Example 8.3. Find i for the following circuit.

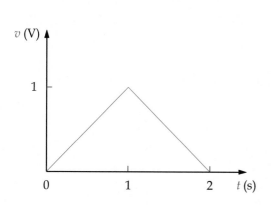

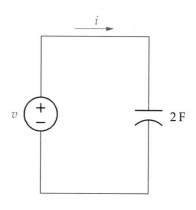

Solution. For $t < 0$ and $t > 2$, $v = 0$ V, and therefore $i = 0$ in this interval. For nonzero values, the voltage waveform is described with two different functions, $v = t$ V for $0 \le t \le 1$, and $v = -(t - 2)$ V for $1 < t \le 2$. Equation (8.2) is used to determine the current for each interval as follows.

For $0 < t < 1$

$$i = C\frac{dv}{dt} = 2 \times \frac{d}{dt}(t) = 2 \text{ A}$$

For $1 \leq t \leq 2$

$$i = C\frac{dv}{dt} = 2 \times \frac{d}{dt}\left(-(t-2)\right) = -2\,\text{A}$$ ∎

Voltage cannot change instantaneously across a capacitor. To have a step change in voltage across a capacitor requires that an infinite current flow through the capacitor, which is not physically possible. Of course, this is mathematically possible using a Dirac delta function.

Equation (8.2) defines the current through a capacitor for a given voltage. Suppose one is given a current through a capacitor and asked to find the voltage. To find the voltage, we start from Eq. (8.2) by multiplying both sides by dt, giving

$$i(t)\,dt = C\,dv$$

Integrating both sides yields

$$\int_{t_0}^{t} i(\lambda)\,d\lambda = C\int_{v(t_0)}^{v(t)} dv$$

or

$$v(t) = \frac{1}{C}\int_{t_0}^{t} i\,dt + v(t_0) \tag{8.4}$$

For $t_0 = 0$, Eq. (8.4) reduces to

$$v(t) = \frac{1}{C}\int_{0}^{t} i\,dt + v(0) \tag{8.5}$$

and for $t_0 = -\infty$, Eq. (8.4) reduces to

$$v(t) = \frac{1}{C}\int_{-\infty}^{t} i\,(\lambda)\,d\lambda \tag{8.6}$$

The initial voltage in Eq. (8.4), $v(t_0)$, is usually defined with the same polarity as v, which means $v(t_0)$ is a positive quantity. If the polarity of $v(t_0)$ is in the opposite direction of v (as will happen when we write mesh equations), then $v(t_0)$ is negative.

Example 8.4.   Find v for the circuit that follows.

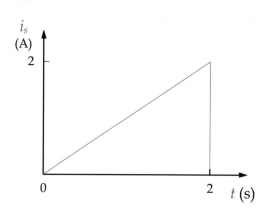

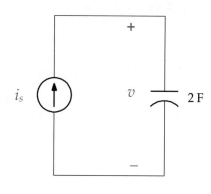

Solution. The current waveform is described with three different functions: for the interval $t \leq 0$, for the interval $0 < t \leq 2$, and for $t > 2$. To find the voltage, we apply Eq. (8.6) for each interval as follows.

For $t < 0$

$$v(t) = \frac{1}{C} \int_{-\infty}^{t} i \, dt = \frac{1}{2} \int_{-\infty}^{0} 0 \, dt = 0 \, \text{V}$$

For $0 \leq t \leq 2$

$$v(t) = \frac{1}{C} \int_{0}^{t} i \, dt + v(0)$$

and with $v(0) = 0$, we have

$$v(t) = \frac{1}{2} \int_{0}^{t} \lambda \, d\lambda = \frac{1}{2} \left(\frac{\lambda^2}{2} \right) \Big|_{0}^{t} = \frac{t^2}{4} \, \text{V}$$

The voltage at $t = 2$ needed for the initial condition in the next part is

$$v(2) = \frac{t^2}{4} \Big|_{t=2} = 1 \, \text{V}$$

For $t > 2$

$$v(t) = \frac{1}{C} \int_{2}^{t} i \, dt + v(2) = \frac{1}{2} \int_{2}^{t} 0 \, dt + v(2) = 1 \, \text{V}$$

■

8.1   POWER AND ENERGY

The capacitor is a passive element that absorbs power according to the relationship

$$p = vi = C v \frac{dv}{dt} \tag{8.7}$$

Energy is stored in the electric field during any period of time $[t_0, t]$ as

$$w(t) - w(t_0) = \int_{t_0}^{t} p \, dt = C \int_{t_0}^{t} v \frac{dv}{dt} \, dt$$

$$= C \int_{v(t_0)}^{v(t)} v \, dv = \frac{1}{2} C([v(t)]^2 - [v(t_0)]^2) \tag{8.8}$$

where the unit of energy is joules (J). At time equal to negative infinity, we assume that the initial voltage is zero and thus the total energy is given by

$$w(t) = \frac{1}{2} C v^2 \tag{8.9}$$

Any energy stored in the electric field is recoverable. If power is negative in Eq. (8.7), energy is being extracted from the capacitor. If power is positive, energy is being stored in the capacitor.

Example 8.5. Find the power and energy for the following circuit.

$$v_s(t) = \begin{cases} 0 \text{ V} & t < 0 \\ t^2 \text{ V} & 0 \leq t \leq 1 \\ e^{-(t-1)} \text{ V} & t > 1 \end{cases}$$

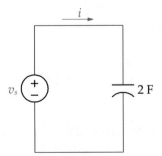

Solution. To find the power, we first need to find the current calculated from Eq. (8.2).

$$i(t) = C \frac{dv}{dt} = \begin{cases} 0 \text{ A} & t < 0 \\ 4t \text{ A} & 0 \leq t \leq 1 \\ -2e^{-(t-1)} \text{ A} & t > 1 \end{cases}$$

Equation (8.7) is used to find the power as follows.

$$p(t) = v \cdot i = \begin{cases} 0 & t < 0 \\ t^2 \cdot 4t = 4t^3 \text{ W} & 0 \le t \le 1 \\ e^{-(t-1)} \cdot (-2e^{-(t-1)}) = -2e^{-2(t-1)} \text{ W} & t > 1 \end{cases}$$

The energy in each interval is calculated using Eq. (8.9) as follows.

$$w(t) = \frac{1}{2}Cv^2 = \begin{cases} 0 & t < 0 \\ \frac{1}{2} \times 2[t^2]^2 = t^4 \text{ J} & 0 \le t \le 1 \\ \frac{1}{2} \times 2[e^{-(t-1)}]^2 = e^{-2(t-1)} \text{ J} & t > 1 \end{cases}$$

Note that energy is being stored in the interval $0 \le t \le 1$ since power is positive, and energy is being extracted from the electric field by the source for $t > 1$ since power is negative.

From Eq. (8.8), we calculate the energy stored in the electric field during time interval $[0, 1]$ as

$$w(1) - w(0) = \frac{1}{2} \times 2(v(1)^2 - v(0)^2) = 1 \text{ J}$$

and the energy delivered from the electric field in the interval $[1, \infty]$ as

$$w(\infty) - w(1) = \frac{1}{2} \times 2(v(\infty)^2 - v(1)^2) = -1 \text{ J} \qquad \blacksquare$$

CHAPTER 9

Inductance and Capacitance Combinations

As with resistors, it is possible to reduce complex inductor and capacitor circuits into simpler, equivalent circuits by combining series and parallel collections of like elements. Consider the following circuit consisting of N inductors in series. Since the same current flows through each inductor, the voltage drop across each inductor is $v_i = L_i \frac{di}{dt}$. Applying KVL on this circuit gives

$$
\begin{aligned}
v_s &= v_1 + v_2 + \cdots + v_N \\
&= L_1 \frac{di}{dt} + L_2 \frac{di}{dt} + \cdots + L_N \frac{di}{dt} \\
&= (L_1 + L_2 + \cdots + L_N) \frac{di}{dt} \\
&= L_{EQ} \frac{di}{dt}
\end{aligned}
$$

Thus, inductors connected in a series can be replaced by an equivalent inductance, whereby

$$
L_{EQ} = L_1 + L_2 + \cdots + L_N \tag{9.1}
$$

Next, consider N inductors connected in parallel as shown in Fig. 9.1. Since the same voltage is across each inductor, the current through each inductor is $i_i = \frac{1}{L_i} \int_{t_0}^{t} v_0 \, dt + i_i(t_0)$. Applying KCL on the upper node of this circuit gives

$$
\begin{aligned}
i_s &= \sum_{i=1}^{N} \left[\frac{1}{L_i} \int_{t_0}^{t} v_0 \, dt + i_i(t_0) \right] \\
&= \left[\frac{1}{L_1} + \frac{1}{L_2} + \cdots + \frac{1}{L_N} \right] \int_{t_0}^{t} v_0 \, dt + \sum_{i=1}^{N} i_i(t_0) \\
&= \frac{1}{L_{EQ}} \int_{t_0}^{t} v_0 \, dt + i_s(t_0)
\end{aligned}
$$

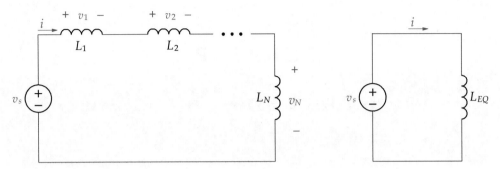

FIGURE 9.1: (Left) N inductors in series. (Right) An equivalent circuit for inductors in series.

Thus, inductors connected in parallel can be replaced by an equivalent inductance, whereby

$$L_{EQ} = \frac{1}{\frac{1}{L_1} + \frac{1}{L_2} + \cdots + \frac{1}{L_N}} \qquad (9.2)$$

For the case of two inductors in parallel, Eq. (9.2) reduces to

$$L_{EQ} = \frac{L_1 \, L_2}{L_1 + L_2} \qquad (9.3)$$

Equations (9.1) and (9.2) are similar to the results we found for resistors in series and parallel, and the process used for simplifying resistor circuits is the same used for inductors in series and parallel.

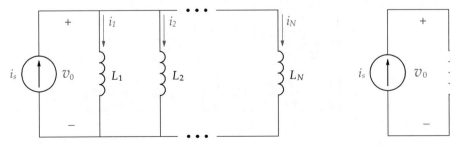

FIGURE 9.2: (Left) N inductors in parallel. (Right) An equivalent circuit for inductors in parallel.

Example 9.1. Find L_{EQ} for the following circuit.

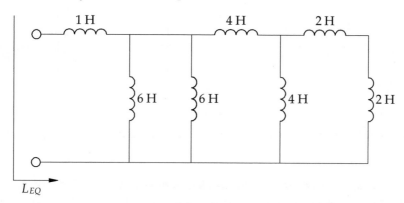

Solution. The equivalent inductance is found by forming series and parallel combinations from right to left until it reduces to a single inductance as shown in the following figure.

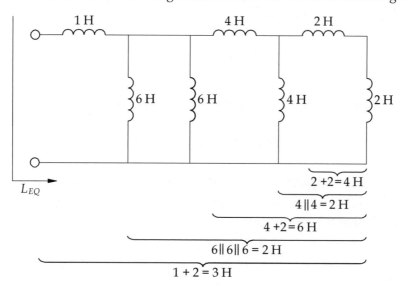

If follows that

$$L_{EQ} = 1 + (6 \parallel 6 \parallel (4 + (4 \parallel (2 + 2))))$$

$$= 1 + \left(6 \parallel 6 \parallel \left(4 + \left(\frac{1}{\frac{1}{4} + \frac{1}{4}}\right)\right)\right)$$

$$= 1 + \left(\frac{1}{\frac{1}{6} + \frac{1}{6} + \frac{1}{6}}\right) = 1 + 2 = 3\,H$$ ∎

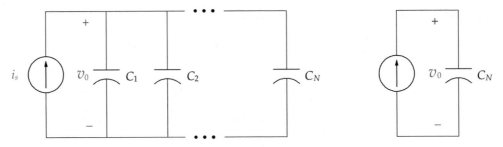

FIGURE 9.3: (Left) N capacitors in parallel. (Right) An equivalent circuit for capacitors in parallel.

Next, consider N capacitors connected in parallel, as shown in Fig. 9.3. Since the same voltage is across each capacitor, the current through each inductor is $i_i = C_i \frac{dv_0}{dt}$. Applying KCL on the upper node of this circuit gives

$$i_s = C_1 \frac{dv_0}{dt} + C_2 \frac{dv_0}{dt} + \cdots + C_N \frac{dv_0}{dt}$$

$$= C_{EQ} \frac{dv_0}{dt}$$

Thus, capacitors connected in series can be replaced by an equivalent capacitance, whereby

$$C_{EQ} = C_1 + C_2 + \cdots + C_N \tag{9.4}$$

Next, consider N capacitors connected in series as shown in Fig. 9.4. Since the same current flows through each capacitor, the voltage across each inductor is $v_i = \frac{1}{C_i} \int_{t_0} i \, dt + v_i(t_0)$. Applying KVL around this circuit gives

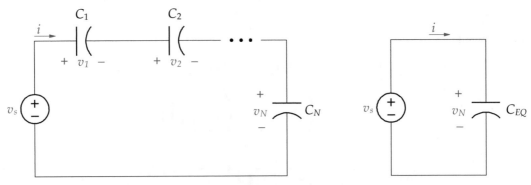

FIGURE 9.4: (Left) N capacitors in series. (Right) An equivalent circuit for capacitors in series.

$$v_s = \sum_{i=1}^{N} \left[\frac{1}{C_i} \int_{t_0}^{t} i \, dt + v_i(t_0) \right]$$

$$= \left[\frac{1}{C_1} + \frac{1}{C_2} + \cdots + \frac{1}{C_N} \right] \int_{t_0}^{t} i \, dt + \sum_{i=1}^{N} v_i(t_0)$$

$$= \frac{1}{C_{EQ}} \int_{t_0}^{t} i \, dt + v_s(t_0)$$

Thus, capacitors connected in series can be replaced by an equivalent capacitance, whereby

$$C_{EQ} = \frac{1}{\frac{1}{C_1} + \frac{1}{C_2} + \cdots + \frac{1}{C_N}} \tag{9.5}$$

For the case of two capacitors in series, Eq. (9.5) reduces to

$$C_{EQ} = \frac{C_1 \, C_2}{C_1 + C_2} \tag{9.6}$$

Equations (9.5) and (9.6) are similar to the results we found for resistors in parallel and series, respectively. That is, we treat capacitors in parallel using the techniques for resistors in series, and capacitors in series as resistors in parallel.

Example 9.2.   Reduce the following circuit to a single capacitor and inductor.

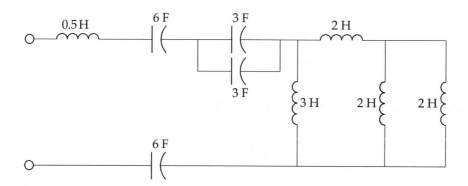

Solution. The equivalent inductance is found by forming series and parallel combinations, from right to left, of inductors and capacitors until the analysis reduces the circuit to a single inductance and capacitance. Consider the inductors first as shown in the following figure on the right, which results in 1.5 H. Also note that the two parallel capacitors equal 6 F.

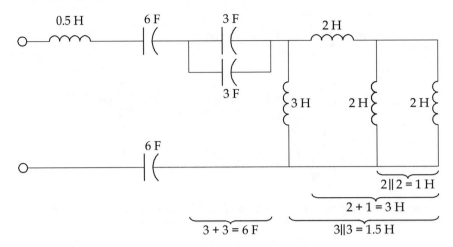

Next we slide the 1.5 H equivalent inductance to the left past the capacitors as shown in the following figure. The two series inductors equal 2 H, and the three capacitors in series (treated like resistors in parallel) equal $\frac{1}{\frac{1}{6}+\frac{1}{6}+\frac{1}{6}} = 2\,\text{F}.$

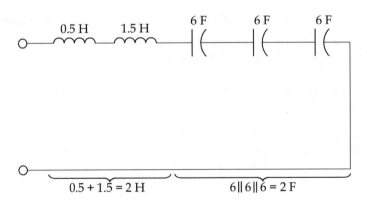

The final reduced circuit is shown in the following figure.

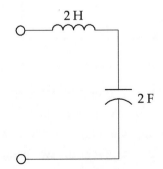

The next example illustrates how to simplify a circuit and then apply the mesh-current method to solve for unknown currents.

Example 9.3. (a) Find $i(t)$, $i_1(t)$, $i_2(t)$ and $v(t)$ for the following circuit for $t \geq 0$ given $i_1(0^-) = 5$ A and $i_2(0^-) = 15$ A. (b) Find the initial energy stored in the inductors. (c) Find the energy dissipated in the resistors between $t = 0$ and $t = \infty$. (d) Find the energy trapped in the inductors at $t = \infty$.

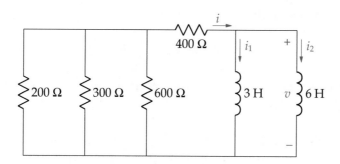

Solution

(a) For ease in solution, first simplify the circuit by combining the three resistors as

$$R_{EQ} = 400 + \frac{1}{\frac{1}{200} + \frac{1}{300} + \frac{1}{600}} = 500 \, \Omega$$

and two inductors as $L_{EQ} = \frac{3 \times 6}{3+6} = 2$ H as shown in the following circuit.

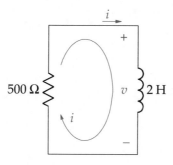

We use the mesh-current method to find $v(t)$ as follows. Recall that the voltage drop across an inductor is $v_L = L\frac{di}{dt}$, so we have

$$500i + 2\frac{di}{dt} = 0$$

or

$$\frac{di}{dt} + 250i = 0$$

The previous differential equation has the characteristic equation

$$s + 250 = 0$$

with root $s = -250$ and solution

$$i(t) = K_1 e^{-250t} \, \text{A}$$

Note that the forced response is zero since there is no input. Because the energy stored in an inductor cannot change instantaneously,

$$i_1(0^-) = i_1(0^+) = 5 \, \text{A}, \, i_2(0^-) = i_2(0^+) = 15 \, \text{A and}$$
$$i(0^-) = i(0^+) = i_1(0^+) + i_2(0^+) = 20 \, \text{A}$$

To determine K_1 we use $i(0) = 20 = K_1$. Thus, our solution is $i(t) = 20e^{-250t} u(t) \, \text{A}$. To find $i_1(t)$ and $i_2(t)$ we find $v(t) = L_{EQ}\frac{di}{dt}$ and then use $i_i(t) = \frac{1}{L_i} \int_0^t v \, dt + i_i(0)$. For $t \geq 0$

$$v(t) = 2\frac{di}{dt} = -10000e^{-250t} u(t) \, \text{V}$$

and

$$i_1(t) = \frac{1}{3} \int_0^t (-10000e^{-250\lambda}) \, d\lambda + 5 = \frac{40}{3} e^{-250\lambda} \Big|_{\lambda=0}^t + 5 = \frac{40}{3} e^{-250t} - \frac{40}{3} + 5$$

$$= \frac{40}{3} e^{-250t} - \frac{25}{3} \, \text{A}$$

$$i_2(t) = \frac{1}{6} \int_0^t (-10000e^{-250\lambda}) \, d\lambda + 15 = \frac{40}{6} e^{-250\lambda} \Big|_{\lambda=0}^t + 15 = \frac{20}{3} e^{-250t} - \frac{20}{3} + \frac{45}{3}$$

$$= \frac{20}{3} e^{-250t} + \frac{25}{3} \, \text{A}$$

Naturally $i(t) = i_1(t) + i_2(t) = 20e^{-250t} u(t) \, \text{A}$.

(b) From Eq. 7.7, the total initial energy stored in the inductors equals

$$w_L(0) = \frac{1}{2}L_1 i_1^2 + \frac{1}{2}L_2 i_2^2 = \frac{1}{2} \times 3 \times 5^2 + \frac{1}{2} \times 6 \times 15^2 = 712.5 \, \text{J}$$

(c) The energy dissipated in the resistors equals

$$w_{R_{EQ}} = \int_0^\infty R_{EQ} i^2(t)\, dt = \int_0^\infty 500 \times (20e^{-250t})^2\, dt$$

$$= \int_0^\infty 500 \times 400 e^{-500t}\, dt = -\left. 400 e^{-500t} \right|_0^\infty = 400\,\mathrm{J}$$

(d) The energy trapped in the inductors at $t = \infty$ equals

$$w_1(\infty) = \frac{1}{2} L_1 i_1^2(\infty) = \frac{1}{2} \times 3 \times \left(-\frac{25}{3}\right)^2 = 104.1667\,\mathrm{J}$$

$$w_2(\infty) = \frac{1}{2} L_2 i_2^2(\infty) = \frac{1}{2} \times 6 \times \left(\frac{25}{3}\right)^2 = 208.333\,\mathrm{J}$$

Notice that energy trapped in the two inductors equals the total initial stored energy minus the energy dissipated in the resistors. Also note that while energy is trapped in the two inductors at $t = \infty$, the energy stored in the equivalent inductor, L_{EQ}, is zero. Note that the initial energy stored in the equivalent inductor is $\frac{1}{2} \times 2 \times (20)^2 = 400\,\mathrm{J}$, and that the energy dissipated in the resistors equals the energy stored in the equivalent inductor at $t = 0$. ∎

CHAPTER 10

A General Approach to Solving Circuits Involving Resistors, Capacitors and Inductors

Sometimes a circuit consisting of resistors, inductors and capacitors cannot be simplified by bringing together like elements in series and parallel combinations. Consider the circuit shown in Fig. 10.1. In this case, the absence of parallel or series combinations of resistors, inductors or capacitors prevents us from simplifying the circuit for ease in solution, as in Ex. 10.1. In this section, we apply the node-voltage and mesh-current methods to write equations involving integrals and differentials using element relationships for resistors, inductors and capacitors. From these equations, we can solve for unknown currents and voltages of interest.

Example 10.1. Write the node equations for the following circuit for $t \geq 0$ if the initial conditions are zero.

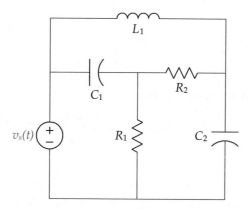

Solution. With the reference node at the bottom of the circuit, we have two essential nodes, as shown in the following redrawn circuit. Recall that the node involving the voltage source is a known voltage and that we do not write a node equation for it. When writing the node-voltage equations, the current through a capacitor is $i_c = C\,\Delta\dot{v}$, where $\Delta\dot{v}$ is the derivative of the

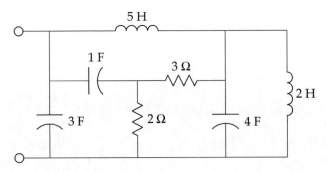

FIGURE 10.1: A circuit that cannot be simplified.

voltage across the capacitor, and the current through an inductor is $i_L = \frac{1}{L} \int_0^t \Delta v d\lambda + i_L(0)$, where Δv is the voltage across the inductor. Since the initial conditions are zero, the term $i_L(0) = 0$.

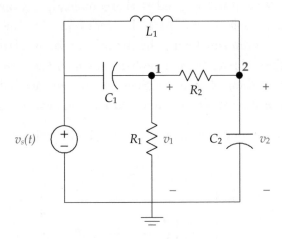

Summing the currents leaving node 1 gives

$$C_1(\dot{v}_1 - \dot{v}_s) + \frac{v_1}{R_1} + \frac{v_1 - v_2}{R_2} = 0$$

which simplifies to

$$C_1\dot{v}_1 + \left(\frac{1}{R_1} + \frac{1}{R_2}\right)v_1 - \frac{1}{R_2}v_2 = C_1\dot{v}_s$$

Summing the currents leaving node 2 gives

$$\frac{v_2 - v_1}{R_2} + C_2\dot{v}_2 + \frac{1}{L_1}\int_0^t (v_2 - v_s)\, d\lambda = 0$$

Typically, we eliminate integrals in the node equations by differentiating. When applied to the previous expression, this gives

$$\frac{1}{R_2}\dot{v}_2 - \frac{1}{R_2}\dot{v}_1 + C_2\ddot{v}_2 + \frac{1}{L_1}v_2 - \frac{1}{L_1}v_s = 0$$

and after rearranging yields

$$\ddot{v}_2 + \frac{1}{C_2 R_2}\dot{v}_2 + \frac{1}{C_2 L_1}v_2 - \frac{1}{C_2 R_2}\dot{v}_1 = \frac{1}{C_2 L_1}v_s \qquad \blacksquare$$

When applying the node-voltage method, we generate one equation for each essential node. To write a single differential equation involving just one node voltage and the inputs, we use the other node equations and substitute into the node equation of the desired node voltage. Sometimes this involves differentiation as well as substitution. The easiest case involves a node equation containing an undesired node voltage without its derivatives. Another method for creating a single differential equation is to use the D operator.

Consider the node equations for Ex. 10.1, and assume that we are interested in obtaining a single differential equation involving node voltage v_1 and its derivatives, and the input. For ease in analysis, let us assume that the values for the circuit elements are $R_1 = R_2 = 1\,\Omega$, $C_1 = C_2 = 1\,\mathrm{F}$, and $L_1 = 1\,\mathrm{H}$, giving us

$$\dot{v}_1 + 2v_1 - v_2 = \dot{v}_s$$

and

$$\ddot{v}_2 + \dot{v}_2 + v_2 - \dot{v}_1 = v_s$$

Using the first equation, we solve for v_2, calculate $\dot{v}_2$ and $\ddot{v}_2$, and then substitute into the second equation as follows.

$$v_2 = \dot{v}_1 + 2v_1 - \dot{v}_s$$
$$\dot{v}_2 = \ddot{v}_1 + 2\dot{v}_1 - \ddot{v}_s$$
$$\ddot{v}_2 = \dddot{v}_1 + 2\ddot{v}_1 - \dddot{v}_s$$

After substituting into the second node equation, we have

$$\dddot{v}_1 + 2\ddot{v}_1 - \dddot{v}_s + \ddot{v}_1 + 2\dot{v}_1 - \ddot{v}_s + \dot{v}_1 + 2v_1 - \dot{v}_s - \dot{v}_1 = v_s$$

and after simplifying

$$\dddot{v}_1 + 3\ddot{v}_1 + 2\dot{v}_1 + 2v_1 = \dddot{v}_s + \ddot{v}_s + \dot{v}_s - v_s$$

The D operator also provides us the means to write the two differential equations as a single differential equation involving only v_1 and v_s. In terms of the D operator, the two node equations are written as

$$(D+2)v_1 - v_2 = Dv_s$$

$$(D^2 + D + 1)v_2 - Dv_1 = v_s$$

Solving the first equation for v_2 gives $v_2 = (D+2) - Dv_s$, and after substituting v_2 into the second equation yields

$$(D^2 + D + 1)(D+2)v_1 - D(D^2 + D + 1)v_s - Dv_1 = v_s$$

Upon simplification

$$(D^3 + 3D^2 + 2D + 2)v_1 = (D^3 + D^2 + D + 1)v_s$$

Returning to the differential notation, the result is

$$\dddot{v}_1 + 3\ddot{v}_1 + 2\dot{v}_1 + 2v_1 = \dddot{v}_s + \ddot{v}_s + \dot{v}_s - v_s$$

which is the same expression we calculated before.

In general, the order of the differential equation relating a single output variable and the inputs is equal to the number of energy storing elements in the circuit (capacitors and inductors). In some circuits, the order of the differential equation is less than the number of capacitors and inductors in the circuit. This occurs when capacitor voltages and inductor currents are not independent, that is, there is an algebraic relationship between the capacitor, specifically voltages and the inputs, or the inductor currents and the inputs. This occurs when capacitors are connected directly to a voltage source or when inductors are connected directly to a current source as shown in Fig. 10.2A and B. In Fig. 10.2A, two capacitors are connected directly to a voltage source. KVL applied around the outer loop gives

$$-v(t) + v_1(t) + v_2(t) = 0$$

or

$$v(t) = v_1(t) + v_2(t)$$

where $v_1(t)$ and $v_2(t)$ are the voltages across the two capacitors. It should be clear that in the previous equation, $v_1(t)$ and $v_2(t)$ are not independent, that is, there is an algebraic relationship

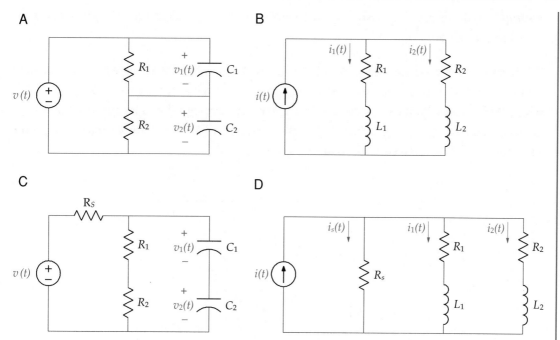

FIGURE 10.2: (A), (B) Circuits described by first-order differential equations and having algebraic relationships between the voltages and currents. (C), (D) Circuits described by second-order differential equations, with no algebraic relationships between voltages and currents.

between the two; if $v_1(t)$ is known, then $v_2(t) = v(t) - v_1(t)$. The same situation occurs with two inductors connected directly to a current source as in Fig. 10.2B. KCL applied at the upper node gives

$$i(t) = i_1(t) + i_2(t)$$

As before, currents $i_1(t)$ and $i_2(t)$ are not independent: there exists an algebraic relationship between the currents. Circuits with the characteristics of Fig. 10.2A and B are rare in realistic situations. For example, the two circuits in Fig. 10.2A and B are better described by those in Fig. 10.2C and D since the voltage and current sources are more appropriately modeled with a resistance within the ideal source, R_s (called an internal resistance—a small resistance for voltage source and a large resistance for a current source). By including an internal resistance R_s in the circuit, we no longer have any algebraic relationships among the voltages and currents.

Notice that the circuit given in Ex. 10.1 has three energy storing elements (two capacitors and one inductor), and the resulting differential equation is of third-order, as expected.

Example 10.2. Write the mesh-current equations for the circuit in Ex. 10.1 for $t \geq 0$ if the initial conditions are zero.

Solution. There are three meshes in this circuit, as shown in the following figure. To write the mesh-current equations, recall that the voltage across a capacitor is $v_C = \frac{1}{C} \int \Delta i \, d\lambda + v_C(0^+)$, where Δi is the resultant current (sum of mesh currents) through the capacitor, and the voltage across a inductor $v_L = L \frac{d\Delta i}{dt}$ where $\frac{d\Delta i}{dt}$ is the derivative of the resultant current through the inductor. Since the initial conditions are zero, $v_C(0^+) = 0$.

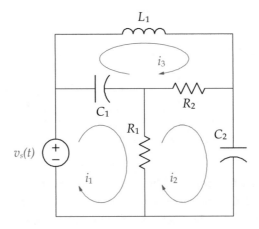

Summing the voltage drops around mesh 1 gives

$$-v_s + \frac{1}{C_1} \int_0^t (i_1 - i_3) \, d\lambda + R_1(i_1 - i_2) = 0$$

Differentiating and rearranging the previous equation gives

$$R_1 \frac{di_1}{dt} + \frac{1}{C_1} i_1 - R_1 \frac{di_2}{dt} + \frac{1}{C_1} i_3 = \frac{dv_s}{dt}$$

Summing the voltage drops around mesh 2 gives

$$R_1(i_2 - i_1) + R_2(i_2 - i_3) + \frac{1}{C_2} \int_0^t i_2 d\lambda = 0$$

Differentiating and rearranging the previous equation gives

$$(R_1 + R_2) \frac{di_2}{dt} + \frac{1}{C_2} i_2 - R_1 \frac{di_1}{dt} - R_2 \frac{di_3}{dt} = 0$$

Finally, summing the voltage drops around mesh 3 gives

$$L_1 \frac{di_3}{dt} + R_2(i_3 - i_2) + \frac{1}{C_1} \int_0^t (i_3 - i_1)d\lambda = 0$$

Differentiating and rearranging the previous equation gives

$$L_1 \frac{d^2 i_3}{dt^2} + R_2 \frac{di_3}{dt} + \frac{1}{C_1} i_3 - R_2 \frac{di_2}{dt} - \frac{1}{C_1} i_1 = 0 \qquad \blacksquare$$

The previous two examples involved a circuit with zero initial conditions. When circuits involve nonzero initial conditions, our approach remains the same as before except that the initial inductor currents are included when writing the node-voltage and the initial capacitor voltages are included when writing the mesh-current equations.

Example 10.3. Write the node equations for the following circuit for $t \geq 0$ assuming the initial conditions are $i_{L_1}(0) = 8\,\text{A}$ and $i_{L_2}(0) = -4\,\text{A}$.

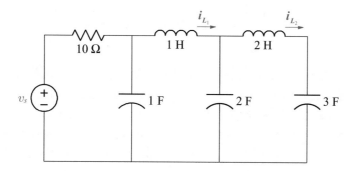

Solution. With the reference node at the bottom of the circuit, we have three essential nodes as shown in the redrawn circuit that follows.

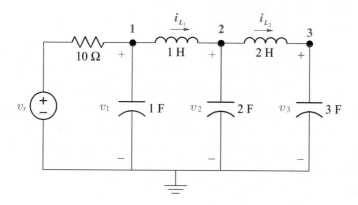

Summing the currents leaving node 1 gives

$$\frac{(v_1 - v_s)}{10} + \dot{v}_1 + \int_0^t (v_1 - v_2)\, d\lambda + 8 = 0$$

where $i_{L_1}(0) = 8\,\text{A}$.

Summing the currents leaving node 2 gives

$$\int_0^t (v_2 - v_1) d\lambda - 8 + 2\dot{v}_2 + \frac{1}{2}\int_0^t (v_2 - v_3) d\lambda - 4 = 0$$

where $i_{L_2}(0) = -4\,\text{A}$. Notice that the sign for the initial inductor current is negative because the direction is from right to left and the current is defined on the circuit diagram in the opposite direction for the node 2 equation.

Summing the currents leaving node 3 gives

$$\frac{1}{2}\int_0^t (v_3 - v_2) d\lambda + 4 + 3\dot{v}_3 = 0$$

In this example, we have not simplified the node equations by differentiating to remove the integral, which would have eliminated the initial inductor currents from the node-equations. If we were to write a single differential equation involving just one node voltage and the input, a fifth-order differential equation would result because there are five energy storing elements in the circuit. To solve the differential equation, we would need five initial conditions, the initial node voltage for the variable selected, as well as the first through fourth derivatives at time zero. ∎

10.1 DISCONTINUITIES AND INITIAL CONDITIONS IN A CIRCUIT

Discontinuities in voltage and current occur when an input such as a unit step is applied or a switch is thrown in a circuit. As we have seen, when solving an nth order differential equation one must know n initial conditions, typically the output variable and its $(n-1)$ derivatives at the time the input is applied or switch thrown. As we will see, if the inputs to a circuit are known for all time, we can solve for initial conditions directly based on energy considerations and not depend on being provided with them in the problem statement. Almost all of our problems involve the input applied at time zero, so our discussion here is focused on time zero, but may be easily extended to any time an input is applied.

Energy cannot change instantaneously for elements that store energy. Thus, there are no discontinuities allowed in current through an inductor or voltage across a capacitor at any time—specifically, the value of the variable remains the same at $t = 0^-$ and $t = 0^+$. In the previous problem when we were given initial conditions for the inductors and capacitors, this implied, $i_{L_1}(0^-) = i_{L_1}(0^+)$ and $i_{L_2}(0^-) = i_{L_2}(0^+)$, and $v_1(0^-) = v_1(0^+)$, $v_2(0^-) = v_2(0^+)$ and $v_3(0^-) = v_3(0^+)$. With the exception of variables associated with current through an inductor and voltage across a capacitor, other variables can have discontinuities, especially at a time when a unit step is applied or when a switch is thrown; however, these variables must obey KVL and KCL.

While it may not seem obvious at first, a discontinuity is allowed for the derivative of the current through an inductor and voltage across a capacitor at $t = 0^-$ and $t = 0^+$ since

$$\frac{di_L(0+)}{dt} = \frac{v_L(0+)}{L} \quad \text{and} \quad \frac{dv_C(0+)}{dt} = \frac{i_C(0+)}{L}$$

as discontinuities are allowed in $v_L(0+)$ and $i_C(0+)$. Keep in mind that the derivatives in the previous expression are evaluated at zero after differentiation, that is

$$\frac{di_L(0+)}{dt} = \frac{di_L(t)}{dt}\bigg|_{t=0+} \quad \text{and} \quad \frac{dv_C(0+)}{dt} = \frac{dv_C(t)}{dt}\bigg|_{t=0+}$$

In calculations to determine the derivatives of variables not associated with current through an inductor and voltage across a capacitor, the derivative of a unit step input may be needed. Here, we assume the derivative of a unit step input is zero at $t = 0^+$.

The initial conditions for variables not associated with current through an inductor and voltage across a capacitor at times of a discontinuity are determined only from the initial conditions from variables associated with current through an inductor and voltage across a capacitor, and any applicable sources. The analysis is done in two steps involving KCL and KVL or using the node-voltage or mesh-current methods.

1. First, we analyze the circuit at $t = 0^-$. Recall that when a circuit is at steady state, an inductor acts as a short circuit and a capacitor acts as an open circuit. Thus at steady state at $t = 0^-$, we replace all inductors by short circuits and capacitors by open circuits in the circuit. We then solve for the appropriate currents and voltages in the circuit to find the currents through the inductors (actually the shorts connecting the sources and resistors) and voltages across the capacitors (actually the open circuits among the sources and resistors).

2. Second, we analyze the circuit at $t = 0^+$. Since the inductor current cannot change in going from $t = 0^-$ to $t = 0^+$, we replace the inductors with current sources whose values are the currents at $t = 0^-$. Moreover, since the capacitor voltage cannot change in

going from $t = 0^-$ to $t = 0^+$, we replace the capacitors with voltage sources whose values are the voltages at $t = 0^-$. From this circuit we solve for all desired initial conditions necessary to solve the differential equation.

Example 10.4. Find $v_C(0^-)$, $v_L(0^-)$, $i_L(0^-)$, $i_{R_1}(0^-)$, $i_{R_2}(0^-)$, $i_{R_3}(0^-)$, $v_C(0^+)$, $v_L(0^+)$, $i_L(0^+)$, $i_{R_1}(0^+)$, $i_{R_2}(0^+)$, $i_{R_3}(0^+)$, and the derivative of each passive element's current and voltage at $t = 0^+$ for the following circuit.

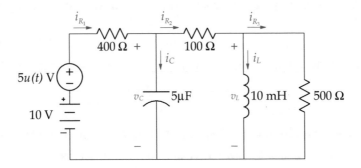

Solution. For $t = 0^-$, the capacitor is replaced by an open circuit and the inductor by a short circuit as shown in the following circuit.

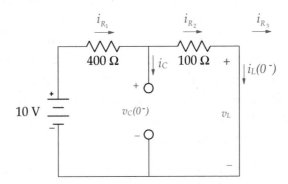

Notice $v_L(0^-) = 0\,\mathrm{V}$ because the inductor is a short circuit. Also note that the $500\,\Omega$ resistor is not shown in the circuit since it is shorted out by the inductor, and so $i_{R_3}(0^-) = 0\,\mathrm{A}$. Using the voltage divider rule, we have

$$v_C(0^-) = 10 \times \frac{100}{400 + 100} = 2\,\mathrm{V}$$

and by Ohm's law

$$i_L(0^-) = \frac{10}{100 + 400} = 0.02\,\mathrm{A}$$

It follows that $i_{R_1}(0^-) = i_{R_2}(0^-) = i_L(0^-) = 0.02$ A. Because voltage across a capacitor and current through an inductor are not allowed to change from $t = 0^-$ to $t = 0^+$ we have $v_C(0^+) = v_C(0^-) = 2$ V and $i_L(0^-) = i_L(0^+) = 0.02$ A.

The circuit for $t = 0^+$ is drawn by replacing the inductors in the original circuit with current sources whose values equal the inductor currents at $t = 0^-$ and the capacitors with voltage sources whose values equal the capacitor voltages at $t = 0^-$ as shown in the following figure with nodes C and L and reference. Note also that the input is now $10 + 5u(t) = 15$ V.

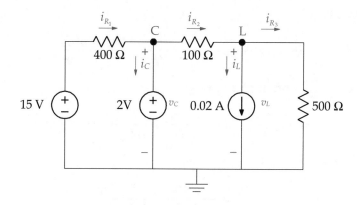

To find $v_L(0^+)$, we sum the currents leaving node L, yielding

$$\frac{v_L - 2}{100} + 0.02 + \frac{v_L}{500} = 0$$

which gives $v_L(0^+) = 0$ V. Now $i_{R_3}(0^+) = \frac{v_L(0^+)}{500} = 0$ A, $i_{R_2}(0^+) = 0.02 + i_{R_3}(0^+) = 0.02$ A, and $i_{R_1}(0^+) = \frac{15-2}{400} = 0.0325$ A.

To find $i_C(0^+)$, we write KCL at node C, giving

$$-i_{R_1}(0^+) + i_C(0^+) + i_{R_2}(0^+) = 0$$

or

$$i_C(0^+) = i_{R_1}(0^+) - i_{R_2}(0^+) = 0.0325 - 0.02 = 0.125 \text{ A}$$

To find $\dot{v}_C(0^+)$, note that $i_C(0^+) = C\dot{v}_C(0^+)$ or

$$\dot{v}_C(0^+) = \frac{i_C(0^+)}{C} = \frac{0.0125}{5 \times 10^{-6}} = 2.5 \times 10^3 \text{ V/s}.$$

Similarly,

$$\frac{di_L(0^+)}{dt} = \frac{v_L(0^+)}{L} = 0 \text{ A/s}$$

Next we have $i_{R_1} = \frac{15 - v_C}{400}$, and

$$\frac{di_{R_1}(0^+)}{dt} = -\frac{1}{400}\frac{dv_C(0^+)}{dt} = -\frac{2.5 \times 10^3}{400} = -6.25 \text{ A/s}$$

To find $\frac{di_{R_3}}{dt}$ we start with KCL $i_{R_3} = i_{R_2} - i_L = i_{R_1} - i_C - i_L$, and

$$\frac{di_{R_3}}{dt} = \frac{di_{R_1}}{dt} - \frac{di_C}{dt} - \frac{di_L}{dt}.$$

For the $\frac{di_C}{dt}$ term we have

$$i_C = i_{R_1} - i_{R_2} \quad \text{and} \quad \frac{di_C}{dt} = \frac{di_{R_1}}{dt} - \frac{di_{R_2}}{dt}$$

Using Ohm's law, the $\frac{di_{R_2}}{dt}$ term is given by as

$$i_{R_2} = \frac{v_C - v_L}{100} \quad \text{and} \quad \frac{di_{R_2}}{dt} = \frac{1}{100}\frac{dv_C}{dt} - \frac{1}{100}\frac{dv_L}{dt}$$

With $v_L = 500 i_{R_3}$ and $\frac{dv_L}{dt} = 500\frac{di_{R_3}}{dt}$, we have

$$\frac{di_{R_2}}{dt} = \frac{1}{100}\frac{dv_C}{dt} - \frac{500}{100}\frac{di_{R_3}}{dt}$$

Substituting $\frac{di_{R_2}}{dt}$ into the $\frac{di_C}{dt}$ equation gives

$$\frac{di_C}{dt} = \frac{di_{R_1}}{dt} - \frac{di_{R_2}}{dt} = \frac{di_{R_1}}{dt} - \frac{1}{100}\frac{dv_C}{dt} + \frac{500}{100}\frac{di_{R_3}}{dt}$$

Returning to the $\frac{di_{R_3}}{dt}$ equation we have

$$\frac{di_{R_3}}{dt} = \frac{di_{R_1}}{dt} - \frac{di_C}{dt} - \frac{di_L}{dt} = \frac{di_{R_1}}{dt} - \frac{di_{R_1}}{dt} + \frac{1}{100}\frac{dv_C}{dt} - \frac{500}{100}\frac{di_{R_3}}{dt} - \frac{di_L}{dt}$$

Collecting the $\frac{di_{R_3}}{dt}$ terms and canceling the $\frac{di_{R_1}}{dt}$ terms, the previous equation reduces to

$$\frac{di_{R_3}}{dt} = \frac{1}{6}\left(\frac{1}{100}\frac{dv_C}{dt} - \frac{di_L}{dt}\right)$$

At $t = 0^+$, we have

$$\frac{di_{R_3}(0^+)}{dt} = \frac{1}{6}\left(\frac{1}{100}\frac{dv_C(0^+)}{dt} - \frac{di_L(0^+)}{dt}\right) = \frac{1}{6}\left(\frac{2.5 \times 10^3}{100} - 0\right) = 4.167 \text{ A/s}$$

From before, we have for $\frac{di_{R_2}}{dt}$

$$\frac{di_{R_2}(0^+)}{dt} = \frac{1}{100}\frac{dv_C(0^+)}{dt} - \frac{500}{100}\frac{di_{R_3}(0^+)}{dt} = \frac{2.5 \times 10^3}{100} - 5 \times 4.167 = 4.167 \text{ A/s}$$

and for $\frac{di_c}{dt}$

$$\frac{di_C(0^+)}{dt} = \frac{di_{R_1}(0^+)}{dt} - \frac{di_{R_2}(0^+)}{dt} = -6.25 - 4.1665 = -10.417 \text{ A/s}$$

Finally, the derivatives of the voltages across the resistors are

$$\frac{dv_{R_1}}{dt} = 400\frac{di_{R_1}}{dt} = 400 \times (-6.25) = -2500 \text{ V/s}$$

$$\frac{dv_{R_2}}{dt} = 100\frac{di_{R_2}}{dt} = 100 \times 4.167 = 416.7 \text{ V/s}$$

and

$$\frac{dv_{R_3}}{dt} = 500\frac{di_{R_3}}{dt} = 500 \times 4.167 = 2083.5 \text{ V/s}$$ ∎

Example 10.5. Find v_C for the circuit in Ex. 10.4 for $t \geq 0$.

Solution. We use the node voltage method to solve this problem since v_C is one of the variables used in the solution. Using the node-voltage method also results in two equations rather than three equations with the mesh-current method. The initial conditions necessary to solve the differential equation for this circuit were calculated in Ex. 10.4. Keep in mind that in Ex. 10.4, we calculated many more initial conditions than are required in this example; here, we only need $v_C(0^+)$ and $\dot{v}_C(0^+)$.

For $t \geq 0$, the circuit is redrawn for analysis in the following figure.

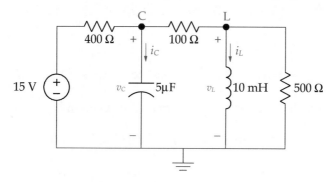

Summing the currents leaving node C gives

$$\frac{v_C - 15}{400} + 5 \times 10^{-6}\dot{v}_C + \frac{v_C - v_L}{100} = 0$$

which simplifies to

$$\dot{v}_C + 2500v_C - 2000v_L = 7500$$

Summing the currents leaving node L gives

$$\frac{v_L - v_C}{100} + \frac{1}{10 \times 10^{-3}} \int_0^t v_L \, d\lambda + i_L(0^+) + \frac{v_L}{500} = 0$$

which, after multiplying by 500 and differentiating, simplifies to

$$6\dot{v}_L + 50 \times 10^3 v_L - 5\dot{v}_C = 0$$

Using the D operator method, our two differential equations are written as

$$Dv_C + 2500v_C - 2000v_L = 7500 \text{ or } (D + 2500)v_C - 2000v_L = 7500$$

$$6Dv_L + 50 \times 10^3 v_L - 5Dv_C = 0 \text{ or } (6D + 50 \times 10^3)v_L - 5Dv_C = 0$$

We then solve for v_L from the first equation,

$$v_L = (0.5 \times 10^{-3} D + 1.25)v_C - 3.75$$

and then substitute v_L into the second equation, giving

$$(6D + 50 \times 10^3)v_L - 5Dv_C = (6D + 50 \times 10^3)((0.5 \times 10^{-3} D + 1.25)v_C - 3.750) - 5Dv_C$$
$$= 0$$

Reducing this expression yields

$$D^2 v_C + 10.417 \times 10^3 Dv_C + 20.83 \times 10^6 v_C = 62.5 \times 10^6$$

Returning to the time domain gives

$$\ddot{v}_C + 10.417 \times 10^3 \dot{v}_C + 20.83 \times 10^6 v_C = 62.5 \times 10^6$$

The characteristic equation for the previous differential equation is

$$s^2 + 10.417 \times 10^3 s + 20.833 \times 10^6 = 0$$

with roots -7.718×10^3 and -2.7×10^3 and the natural solution

$$v_{C_n}(t) = K_1 e^{-7.718 \times 10^3 t} + K_2 e^{-2.7 \times 10^3 t} \text{ V}$$

Next, we solve for the forced response, assuming that $v_{C_f}(t) = K_3$. After substituting into the differential equation, this gives

$$20.833 \times 10^6 K_3 = 62.5 \times 10^6$$

or $K_3 = 3$. Thus, our solution is now

$$v_C(t) = v_{C_n}(t) + v_{C_f}(t) = K_1 e^{-7.718 \times 10^3 t} + K_2 e^{-2.7 \times 10^3 t} + 3\,\text{V}$$

We use the initial conditions to solve for K_1 and K_2. First

$$v_C(0) = 2 = K_1 + K_2 + 3$$

Next

$$\dot{v}_C(t) = -7.718 \times 10^3 K_1 e^{-7.718 \times 10^3 t} - 2.7 \times 10^3 K_2 e^{-2.7 \times 10^3 t}$$

and at $t = 0$,

$$\dot{v}_C(0) = 2.5 \times 10^3 = -7.718 \times 10^3 K_1 - 2.7 \times 10^3 K_2$$

Solving gives $K_1 = 0.04$ and $K_2 = -1.04$. Substituting these values into the solution gives

$$v_C(t) = 0.04 e^{-7.718 \times 10^3 t} - 1.04 e^{-2.7 \times 10^3 t} + 3\,\text{V}$$

for $t \geq 0$.   ∎

Our approach remains the same for circuits with controlled sources. We analyze the circuit for $t = 0^-$ and $t = 0^+$ to establish the initial conditions with the controlled source operational. If the voltage or current used in the controlled source is zero, we replace it with an open circuit if it is a current source and a short circuit if it is a voltage source. If a circuit has an independent voltage or current source that is introduced at zero via a unit step function, for $t = 0^-$ we replace the independent source by a short circuit if it is a voltage source and an open circuit if it is a current source.

Example 10.6. Find i_{L_1} for the following circuit for $t \geq 0$ using the mesh-current method.

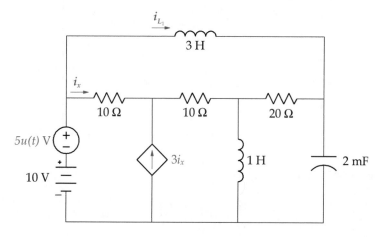

Solution. To solve this problem, we will use the mesh-current method with currents defined in the following circuit. The steps involve:

1. Eliminate all currents except i_4, which is i_{L_1}, which gives a third-order differential equation

2. Solve for the initial conditions

3. Solve the differential equation.

It should be clear that a third-order differential equation should describe this system, since there are three energy storing elements.

Mesh Equations

To write the mesh-current equations, recall that the voltage across a capacitor is $v_C = \frac{1}{C} \int_0^t \Delta i \, d\lambda + v_C(0^+)$, where Δi is the resultant current (sum of mesh currents) through the inductor, and the voltage across a inductor $v_L = L \frac{d\Delta i}{dt}$ where $\frac{d\Delta i}{dt}$ is the derivative of the resultant current through the inductor. Since there is a controlled current source in the circuit, we form a supermesh for meshes 1 and 2. Notice that mesh current $i_4 = i_{L_1}$.

Summing the voltages around supermesh $1 + 2$ yields

$$-15 + 10(i_1 - i_4) + 10(i_2 - i_4) + \frac{d(i_2 - i_3)}{dt} = 0$$

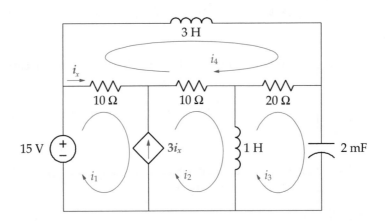

Rearranging the previous equation gives

$$10i_1 + \frac{di_2}{dt} + 10i_2 - \frac{di_3}{dt} - 20i_4 = 15$$

Summing the voltage drops around mesh 3 yields

$$\frac{d(i_2 - i_3)}{dt} + 20(i_3 - i_4) + \frac{1}{2 \times 10^{-3}} \int_0^t i_3 \, d\lambda + v_C(0^+) = 0$$

Differentiating and rearranging the previous equation reduces to

$$-\frac{d^2 i_2}{dt^2} + \frac{d^2 i_3}{dt^2} + 20\frac{di_3}{dt} + 500 i_3 - 20\frac{di_4}{dt} = 0$$

Summing the voltage drops around mesh 4 gives

$$3\frac{di_4}{dt} + 20(i_4 - i_3) + 10(i_4 - i_2) + 10(i_4 - i_1) = 0$$

and after rearranging, we find

$$-10i_1 - 10i_2 - 20i_3 + 3\frac{di_4}{dt} + 40i_4 = 0$$

Applying KCL for the dependent current source gives

$$3i_x = i_2 - i_1$$

Since $i_x = 3(i_1 - i_4)$, we have

$$4i_1 - i_2 - 3i_4 = 0$$

At this time, we have four equations and four unknowns. The dependent current source equation, an algebraic equation, is used to eliminate i_1 from the three mesh equations, since $i_1 = \frac{1}{4}(i_2 + 3i_4)$. Substituting i_1 into the supermesh $1 + 2$ equation and collecting like terms gives

$$\frac{di_2}{dt} + \frac{50}{4}i_2 - \frac{di_3}{dt} - \frac{50}{4}i_4 = 15$$

The mesh 3 equation did not have an i_1, but is written here for convenience.

$$-\frac{d^2 i_2}{dt^2} + \frac{d^2 i_3}{dt^2} + 20\frac{di_3}{dt} + 500 i_3 - 20\frac{di_4}{dt} = 0$$

After substituting i_1 and collecting like terms in the mesh 4 equation, we have

$$-\frac{50}{4}i_2 - 20i_3 + 3\frac{di_4}{dt} + \frac{130}{4}i_4 = 0$$

The mesh 4 equation is used to eliminate i_3 from the other two equations. Solving for i_3 and taking its derivatives yields

$$i_3 = \frac{1}{20}\left(-\frac{50}{4}i_2 + 3\frac{di_4}{dt} + \frac{130}{4}i_4\right)$$

$$\frac{di_3}{dt} = \frac{1}{20}\left(-\frac{50}{4}\frac{di_2}{dt} + 3\frac{d^2i_4}{dt^2} + \frac{130}{4}\frac{di_4}{dt}\right)$$

$$\frac{d^2i_3}{dt^2} = \frac{1}{20}\left(-\frac{50}{4}\frac{d^2i_2}{dt^2} + 3\frac{d^3i_4}{dt^3} + \frac{130}{4}\frac{d^2i_4}{dt^2}\right)$$

Substituting the $\frac{di_3}{dt}$ term into the supermesh $1 + 2$ equation and simplifying gives

$$\frac{13}{8}\frac{di_2}{dt} + \frac{50}{4}i_2 - \frac{3}{20}\frac{d^2i_4}{dt^2} - \frac{13}{8}\frac{di_4}{dt} - \frac{50}{4}i_4 = 15$$

Substituting i_3, $\frac{di_3}{dt}$ and $\frac{d^2i_3}{dt^2}$ into the mesh 3 equation after collecting like terms gives

$$-\frac{13}{8}\frac{d^2i_2}{dt^2} - \frac{50}{4}\frac{di_2}{dt} - 312.5i_2 + \frac{3}{20}\frac{d^3i_4}{dt^3} + \frac{37}{8}\frac{d^2i_4}{dt^2} + \frac{350}{4}\frac{di_4}{dt} + 812.5i_4 = 0$$

To eliminate i_2, we use the D operator method on our two differential equations giving

$$\left(\frac{13}{8}D + \frac{50}{4}\right)i_2 + \left(-\frac{3}{20}D^2 - \frac{13}{8}D - \frac{50}{4}\right)i_4 = 15$$

$$\left(-\frac{13}{8}D^2 - \frac{50}{4}D - 312.5\right)i_2 + \left(\frac{3}{20}D^3 + \frac{37}{8}D^2 + \frac{350}{4}D + 812.5\right)i_4 = 0$$

To solve for i_4, we pre-multiply the first equation by $(-\frac{13}{8}D^2 - \frac{50}{4}D - 312.5)$, pre-multiply the second equation by $(\frac{13}{8}D + \frac{50}{4})$, and then subtract the first equation from the second equation. This yields

$$(4.875D^3 + 112.5D^2 + 1750D + 6250)i_4 = -(-\frac{13}{8}D^2 - \frac{50}{4}D - 312.5) \times 15$$

Returning to the time domain gives

$$4.875\frac{d^3i_4}{dt^3} + 112.5\frac{d^2i_4}{dt^2} + 1750\frac{di_4}{dt} + 6250i_4 = 4687.5$$

Note that the derivative terms on the right-hand side of the D operator equation are zero since the derivative of a constant is zero. Dividing the previous differential equation by 4.875 to put it in a more convenient form results in

$$\frac{d^3 i_4}{dt^3} + 23.1 \frac{d^2 i_4}{dt^2} + 359 \frac{d i_4}{dt} + 6250 i_4 = 1282.1$$

Initial Conditions

The next step in the solution involves finding the initial conditions for the circuit necessary to solve the differential equation, $i_{L_1}(0^+)$, $\frac{d i_{L_1}(0^+)}{dt}$ and $\frac{d^2 i_{L_1}(0^+)}{dt^2}$. As before, this involves:

1. Analyzing the circuit at $t = 0^-$ with the inductors replaced as short circuits and the capacitor as an open circuit

2. Analyzing the circuit at $t = 0^+$ with inductors replaced as current sources whose values equal the inductor currents at $t = 0^-$ and the capacitor as a voltage source whose value equal the capacitor voltage at $t = 0^-$.

For $t = 0^-$, for a node-voltage solution the circuit is redrawn as shown in the following diagram.

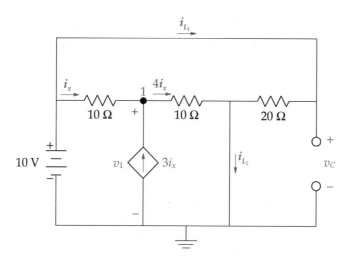

Summing the currents leaving node 1 gives

$$\frac{v_1 - 10}{10} - 3 \left(\frac{10 - v_1}{10} \right) + \frac{v_1}{10} = 0$$

where $i_x = \frac{10 - v_1}{10}$. Solving this equation yields $v_1(0^-) = 8\,\text{V}$. Ohm's law gives $i_{L_1}(0^-) = \frac{10}{20} = 0.5\,\text{A}$ and $i_x(0^-) = \frac{10 - v_1(0^-)}{10} = \frac{10 - 8}{10} = 0.2\,\text{A}$. Since the capacitor is connected across the battery, $v_C(0^-) = 10\,\text{V}$. KCL gives the current through the other inductor as $i_{L_2}(0^-) = 4 i_x(0^-) + i_{L_1}(0^-) = 0.8 + 0.5 = 1.3\,\text{A}$.

For $t = 0^+$, for a node-voltage solution the circuit is redrawn as shown in the following diagram.

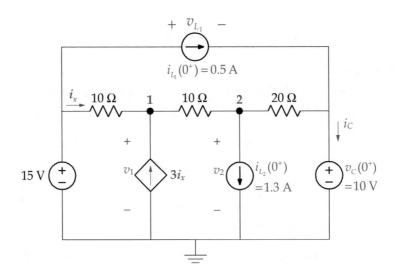

To find the initial conditions for the circuit at $t = 0^+$, we first solve for the node-voltages $v_1(0^+)$ and $v_2(0^+)$ and then solve for $\frac{di_{L_1}(0^+)}{dt}$ and $\frac{d^2i_{L_1}(0^+)}{dt^2}$. Summing the currents leaving node 1 yields

$$\frac{v_1 - 15}{10} - 3\left(\frac{15 - v_1}{10}\right) + \frac{v_1 - v_2}{10} = 0$$

Simplifying the equation, yields

$$5v_1 - v_2 = 60$$

Summing the currents leaving node 2 yields

$$\frac{v_2 - v_1}{10} + 1.3 + \frac{v_2 - 10}{20} = 0$$

Simplifying the equation, we have

$$-2v_1 + 3v_2 = -16$$

Solving the two simultaneous node equations gives

$$v_1(0^+) = 12.6154 \quad \text{and} \quad v_2 0^+) = 3.0769.$$

Recall that $v_C(0^+) = v_C(0^-) = 10\,\text{V}$, $i_{L_1}(0^+) = i_{L_1}(0^-) = 0.5\,\text{A}$, and $i_{L_2}(0^+) = i_{L_2}(0^-) = 1.3\,\text{A}$ from the analysis at $t = 0^-$.

Our next tasks involve finding $\frac{di_{L_1}(0^+)}{dt}$ and $\frac{d^2 i_{L_1}(0^+)}{dt^2}$ using the relationships $v_L = L_1 \frac{di_{L_1}}{dt}$ and $i_C = C \frac{dv_C}{dt}$. Thus

$$\frac{di_{L_1}(0^+)}{dt} = \frac{1}{3} v_{L_1}(0^+)$$

and using KVL to find $v_{L_1}(0^+)$ gives

$$-15 + v_{L_1}(0^+) + v_C(0^+) = 0$$

or

$$v_{L_1}(0^+) = 15 - v_C(0^+) = 15 - 10 = 5 \text{ V}$$

and

$$\frac{di_{L_1}(0^+)}{dt} = \frac{1}{3} v_{L_1}(0^+) = \frac{5}{3} \frac{\text{A}}{\text{s}}$$

We also have $\frac{d^2 i_{L_1}(0^+)}{dt^2} = \frac{1}{3} \frac{dv_{L_1}(0^+)}{dt}$, and from KVL before $v_{L_1} = 15 - v_C$ and after differentiating $\frac{dv_{L_1}}{dt} = 0 - \frac{dv_C}{dt} = -\frac{dv_C}{dt}$. Thus $\frac{d^2 i_{L_1}(0^+)}{dt^2} = \frac{1}{3} \frac{dv_{L_1}(0^+)}{dt} = -\frac{1}{3} \frac{dv_C(0^+)}{dt}$. Now

$$i_C(0^+) = \frac{v_2(0^+) - 10}{20} + i_{L_1}(0^+) = \frac{3.0796 - 10}{20} + 0.5 = 0.154 \text{ A}$$

and

$$\frac{dv_C(0^+)}{dt} = \frac{1}{C} i_C(0^+) = \frac{1000}{2} \times 0.154 = 77 \text{ V/s}$$

Therefore

$$\frac{d^2 i_{L_1}(0^+)}{dt^2} = -\frac{1}{3} \frac{dv_C(0^+)}{dt} = -25.6667 \text{ A}^2/\text{s}.$$

Solving the Differential Equation

The differential equation describing the circuit is

$$\frac{d^3 i_4}{dt^3} + 23.1 \frac{d^2 i_4}{dt^2} + 359 \frac{di_4}{dt} + 6250 i_4 = 1282.1$$

As before, the natural solution is determined first by finding the roots of characteristic equation,

$$s^3 + 23.1 s^2 + 359 s + 6250$$

as $s_1 = -20.4749$, $s_{2,3} = -1.3125 \pm j17.4221$. The roots give rise to the natural response

$$i_{4_n}(t) = K_1 e^{-20.4749t} + e^{-1.3125t}(K_2 \cos(17.4221t) + K_3 \sin(17.4221t))$$

Next we solve for the forced response to the input by assuming

$$i_{4_f}(t) = K_4$$

which, when substituted into the original differential equation, yields

$$6250 K_4 = 1281.1$$

giving $K_4 = 0.2050$. The total solution equals the natural and forced response, written as

$$i_4(t) = i_{4_n}(t) + i_{4_f}(t)$$
$$= K_1 e^{-20.4749t} + e^{-1.3125t}(K_2 \cos(17.4221t) + K_3 \sin(17.4221t)) + 0.2050$$

The initial conditions are used to determine the constants K_1, K_2 and K_3. From the solution for $i_4(t)$, we have

$$i_4(0) = \frac{1}{2} = K_1 + K_2 + 0.2050$$

To use the next initial condition, we find the derivative of the solution, giving

$$\frac{di_4}{dt} = -20.4749 K_1 e^{-20.4749t} - 1.3125 e^{-1.3125t}(K_2 \cos(17.4221t) + K_3 \sin(17.4221t))$$
$$+ e^{-1.3125t}(-17.4221 K_2 \sin(17.4221t) + 17.4221 K_3 \cos(17.4221t))$$

We evaluate this expression at $t = 0$, giving

$$\frac{di_4(0)}{dt} = \frac{5}{3} = -20.4749 K_1 - 1.3125 K_2 + 17.4221 K_3$$

Finally, we take the second derivative of the solution giving

$$\frac{d^2 i_4}{dt^2} = 419.2215 K_1 e^{-20.4749t} + 1.7227 e^{-1.3125t}(K_2 \cos(17.4221t) + K_3 \sin(17.4221t))$$
$$- 1.3125 e^{-1.3125t}(-17.4221 K_2 \sin(17.4221t) + 17.4221 K_3 \cos(17.4221t))$$
$$- 1.3125 e^{-1.3125t}(-17.4221 K_2 \sin(17.4221t) + 17.4221 K_3 \cos(17.4221t))$$
$$+ e^{-1.3125t}(-303.5296 K_2 \cos(17.4221t) - 303.5296 K_3 \sin(17.4221t))$$

We evaluate this expression at $t = 0$, giving

$$\frac{d^2 i_4(0)}{dt^2} = -25.6667 = 419.2215 K_1 + 1.7227 K_2 - 2.6250 \times 17.4221 K_3 - 303.5296 K_2$$
$$= 419.2215 K_1 - 301.807 K_2 - 45.733 K_3$$

The three initial condition equations are put into matrix form for a straightforward solution as $AK = F$ or

$$\begin{bmatrix} 1 & 1 & 0 \\ -20.4749 & -1.3125 & 17.4221 \\ 419.2215 & -301.807 & -45.733 \end{bmatrix} \begin{bmatrix} K_1 \\ K_2 \\ K_3 \end{bmatrix} = \begin{bmatrix} 0.295 \\ \dfrac{5}{3} \\ -25.6667 \end{bmatrix}$$

Using MATLAB, we have

$\gg$ A = [1 1 0; $-$20.4749 $-$1.3125 17.4221; 419.2215 $-$301.807 $-$ 45.733];

$\gg$ F = [0.295; 5/3; $-$25.6667];

$\gg$ K = A\F

K =

0.1025

0.1925

0.2306

The complete solution is

$$i_{L_1}(t) = i_4(t) = 0.1025e^{-20.4749t} + e^{-1.3125t}(0.1925\cos(17.4221t)$$
$$+ 0.2306\sin(17.4221t)) + 0.2050$$

for $t \geq 0$.   ∎

10.2 CIRCUITS WITH SWITCHES

To finish this section, we consider circuits with multiple switches. In effect, each time a switch is thrown, we solve the circuit using the techniques of this section. That is, at each switch time, t_i, we determine the necessary voltages and currents at t_i^- and t_i^+ and solve the circuit problem, move to the next switch time, determine the necessary voltages and currents at t_{i+1}^- and t_{i+1}^+ and solve the circuit problem, and so on. For ease in solving the circuit after the first switch time at $t = 0$, we replace the variable t with $t - t_1$ where t_1 is the switch time and use the unit step function, and repeat this substitution at each switch time. Each time interval is separately analyzed, with the only carry-over from one time interval to the next being the voltages across the capacitor and currents through the inductors at the switch time. The next example illustrates this approach.

Example 10.7. Find v_C for the following circuit for $t \geq 0$.

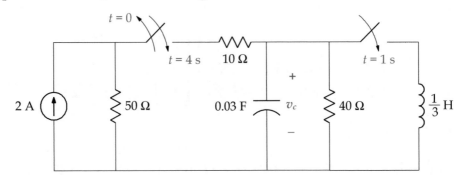

Solution. There are two switches operating in this circuit, with switching times at $t = 0$, $t = 1$ and $t = 4\,s$. We therefore break up the solution into three time intervals, $0 \leq t < 1$, $1 \leq t < 4$ and $t \geq 4\,s$, while realizing that we need to know the initial conditions just before and after each switch time, and that we also need to solve for the steady-state solution for $t < 0$.

For $t < 0$

At steady state, the capacitor is replaced by an open circuit as shown in the following figure.

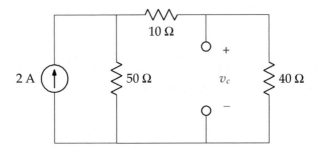

Using the current divider rule and Ohm's law, we have $v_C(0^-) = 40\,\text{V}$. Note that voltage cannot instantaneously change across a capacitor, so $v_C(0^+) = 40\,\text{V}$.

For $0 \leq t < 1$

During this interval, the switch on the left opens, leaving us with the following circuit to analyze.

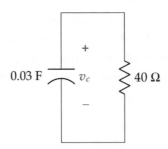

Notice the 10 Ω resistor is eliminated since no current flows through it because it is an open circuit. Using KCL we have

$$0.03\dot{v}_C + \frac{v_C}{40} = 0$$

The differential equation has a root equal to $-\frac{1}{1.2}$ and a solution in this time interval of

$$v_C = v_C(0^+)e^{-\frac{t}{1.2}}u(t) = 40e^{-\frac{t}{1.2}}u(t)\,\text{V}$$

For $1 \le t < 4$

When $t = 1$, the switch on the right closes, introducing an inductor as shown in the following figure.

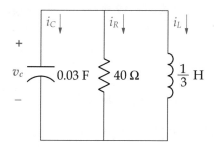

Applying KCL yields

$$0.03\frac{dv_C}{dt} + \frac{v_C}{40} + 5\int_1^t v_C\,d\lambda + i_L(1^+) = 0$$

Dividing the previous equation by 0.03 and differentiating gives

$$\ddot{v}_C + \frac{1}{1.2}\dot{v}_C + \frac{5}{0.03}v_C = 0$$

with complex roots determined from MATLAB as $-0.42 \pm 12.9j$. The solution consists of the natural solution only, since the forced response is zero, and is given by

$$v_C = e^{-0.42(t-1)}(K_1\cos(12.9(t-1)) + K_2\sin(12.9(t-1)))\,(u(t-1) - u(t-4))$$

This solution requires two initial conditions. Recall that at $t = 1$, $v_C(1^+) = v_C(1^-)$ and $i_L(1^+) = i_L(1^-)$. Moreover, all other voltages and currents and their derivatives change from $t = 1^-$ to $t = 1^+$ s. Now using the solution for v_C from the previous interval, we have

$$v_C(1^+) = v_C(1^-) = 40e^{-\frac{1}{1.2}} = 17.4\,\text{V}$$

From our solution in this interval, we have

$$v_C(1^+) = K_1 = 17.4\,\text{V}$$

Since no current was flowing through the inductor during the interval $0 \le t < 1$, $i_L(1^+) = i_L(1^-) = 0\,\text{A}$. Now $i_R(1^+) = \frac{v_C(1^+)}{40} = \frac{17.4}{40} = 0.44\,\text{A}$, and accordingly

$$i_C(1^+) = -(i_L(1^+) + i_R(1^+)) = -(0 + 0.44) = -0.44\,\text{A}$$
$$\dot{v}_C(1^+) = \frac{1}{C}i_C(1^+) = -14.5\,\text{V}$$

Now

$$\dot{v}_C = \begin{pmatrix} -0.42e^{-0.42(t-1)}(K_1 \cos(12.9(t-1)) + K_2 \sin(12.9(t-1))) \\ + e^{-0.42(t-1)}(-12.9K_1 \sin(12.9(t-1)) + 12.9K_2 \cos(12.9(t-1))) \end{pmatrix} \\ \times (u(t-1) - u(t-4))$$

and

$$\dot{v}_C(1^+) = -14.5 = -0.42K_1 + 12.9K_2 = -0.42 \times 17.4 + 12.9K_2$$

which gives $K_2 = -0.557$. Our complete solution is

$$v_C = e^{-0.42(t-1)}(17.4\cos(12.9(t-1)) - 0.557\sin(12.9(t-1)))(u(t-1) - u(t-4))\,\text{V}$$

For $t > 4$.

For the last interval, the switch on the left closes, giving us the following figure to analyze.

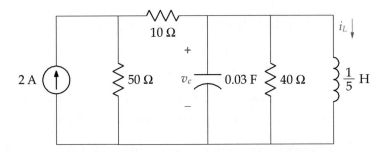

To ease the solution, the 2 A current source with 50 Ω resistor is transformed into a 100 V voltage source in series with a 50 Ω resistor. The 50 and 10 Ω resistors are summed together, and the 100 V voltage source and 60 Ω resistor are transformed into a $\frac{10}{6}$ A current source in parallel with a 60 Ω resistor. The 60 and 40 Ω resistors are combined to give a 24 Ω resistor. The reduced circuit is shown in the following circuit.

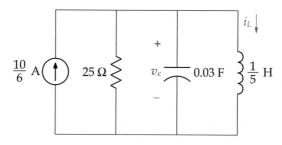

Applying KCL gives

$$-\frac{10}{6} + \frac{v_C}{24} + 0.03\dot{v}_C + 5\int_1^t v_C\,d\lambda + i_L(4^+) = 0$$

Differentiating the previous equation and dividing by 0.03 yields

$$\ddot{v}_C + 1.4\dot{v}_C + 166.7v_C = 55.6$$

with roots determined from MatLab as $-0.7 \pm 12.9j$. The roots are complex, which gives rise to the natural response

$$v_C = e^{-0.7(t-4)}(K_1 \cos(12.9(t - 4)) + K_2 \sin(12.9(t - 4)))\,u(t - 4)$$

Next, we solve for the forced response to the input by assuming $y_f(t) = K_3$. When substituted into the differential equation describing this interval, this yields

$$166.7K_3 = 55.6$$

giving 11.1. The total solution in this interval equals the natural and forced response, written as

$$v_C = 11.1 + e^{-0.7(t-4)}(K_1 \cos(12.9(t - 4)) + K_2 \sin(12.9(t - 4)))\,u(t - 4)$$

Two initial conditions ($v_C(4^+)$ and $\dot{v}_C(4^+)$) are needed to determine the constants K_1 and K_2. As before, at $t = 4$, $v_C(4^+) = v_C(4^-)$ and $i_L(4^+) = i_L(4^-)$, and all other voltages and currents and their derivatives change from $t = 4^-$ to $t = 4^+$ s. Using the solution for v_C from the previous interval, we have

$$v_C(4^+) = v_C(4^-) = e^{-0.42(t-1)}(17.4\cos(12.9(t - 1)) - 0.557\sin(12.9(t - 1)))$$
$$\times (u(t - 1) - u(t - 4))|_{t=4^-}$$
$$= e^{-0.42(4-1)}(17.4\cos(12.9(4 - 1)) - 0.557\sin(12.9(4 - 1)))$$
$$= 2.5\ \text{V}$$

From our solution in this interval, we have

$$v_C(4^+) = 11.1 + K_1 = 2.53 \, \text{V}$$

and $K_1 = -8.6 \, \text{V}$. The next initial condition, $\dot{v}_C(4^+) = \frac{1}{0.03} i_C(4^+)$, is determined from KCL

$$-\frac{10}{6} + \frac{v_C(4^+)}{24} + i_C(4^+) + i_L(4^+) = 0$$

where $i_L(4^+) = i_L(4^-)$. To find $i_L(4^-)$ we use the solution for v_C in the previous interval and compute

$$i_L(4^+) = i_L(4^-) = 5 \int_1^4 v_C \, dt + i_L(1^+) = 5 \int_1^4 v_C \, dt + 0$$

$$= 5 \int_1^4 \left(e^{-0.42(t-1)}(17.4 \cos(12.9(t-1)) - 0.557 \sin(12.9(t-1))) \right) dt$$

$$= \left. e^{-0.42(t-1)}(-0.0037 \cos(12.9(t-1)) + 6.74 \sin(12.9(t-1))) \right|_1^4$$

$$= 1.614 \, \text{A}$$

Now $i_R(1^+) = \frac{v_C(4^+)}{24} = \frac{2.53}{24} = 0.105 \, \text{A}$, and

$$i_C(4^+) = \frac{10}{6} - \frac{v_C(4^+)}{24} - i_L(4^+) = \frac{10}{6} - \frac{2.53}{24} - 1.614 = -0.05 \, \text{A}$$

$$\dot{v}_C(4^+) = \frac{1}{C} i_C(4^+) = -1.75 \, \text{V}$$

Now

$$\dot{v}_C = \begin{pmatrix} -0.7 e^{-0.7(t-4)} \left(K_1 \cos(12.9(t-4)) + K_2 \sin(12.9(t-4)) \right) \\ + e^{-0.7(t-4)} \left(-12.9 K_1 \sin(12.9(t-4)) + 12.9 K_2 \cos(12.9(t-4)) \right) \end{pmatrix} u(t-4)$$

and

$$\dot{v}_C(4^+) = -1.75 = -0.7 K_1 + 12.9 K_2$$

with $K_1 = -8.6$, we have $K_2 = -0.6$. Our solution for this interval is

$$v_C = 11.1 + e^{-0.7(t-4)}(-8.6 \cos(12.9(t-4)) - 0.6 \sin(12.9(t-4))) \, u(t-4) \, \text{V}$$

Summarizing, our complete solution is

$$v_C = \begin{cases} 40e^{-\frac{t}{1.2}} \text{ V} & 0 \leq t < 1 \\ e^{-0.42(t-1)}\left(17.4\cos(12.9\,(t-1)) - 0.557\sin(12.9\,(t-1))\right) \text{ V} & 1 \leq t < 4 \\ 11.1 + e^{-0.7(t-4)}(-8.6\cos(12.9(t-4)) - 0.6\sin(12.9(t-4))) \text{ V} & t \geq 4 \end{cases}$$

■

CHAPTER 11

Operational Amplifiers

In Section 3, we considered controlled voltage and current sources that are dependent on a voltage or current elsewhere in a circuit. These devices were modeled as a two terminal device. Here we consider the operational amplifier, also known as an op amp, a multi-terminal device. An operational amplifier is an electronic device that consists of large numbers of transistors, resistors and capacitors—to fully understand its operation requires knowledge of diodes and transistors, topics not covered in this book. However, to appreciate how an operational amplifier operates in a circuit involves a topic already covered, the controlled voltage source.

As the name implies, the operation amplifier is an amplifier, but when combined with other circuit element, it integrates, differentiates, sums and subtracts. One of the first operational amplifiers approved as an eight-lead dual-in-line package (DIP) is shown in Fig. 11.1. Differing from previous circuit elements, this device has two inputs and one output terminals. Rather than draw the operational amplifier using Fig 11.1, the operational amplifier is drawn with the symbols in Fig. 11.2. The input terminals are labeled the noninverting input (+) and the inverting input (−). The power supply terminals are labeled V+ and V−, which are frequently omitted since they do not affect the circuit behavior except in saturation conditions as will be described. Most people shorten the name of the operational amplifier to the "op amp".

Illustrated in Fig. 11.3 is a model of the op amp focusing on the internal behavior of the input and output terminals. The input–output relationship is

$$v_o = A(v_p - v_n) \tag{11.1}$$

Since the internal resistance is very large, we will replace it with an open circuit to simplify analysis leaving us with the op amp model show in Fig. 11.4.

With the replacement of the internal resistance with an open circuit, the currents $i_n = i_p = 0\,\text{A}$. In addition, current i_A, the current flowing out of the op amp, is not zero. Because i_A is unknown, we seldom apply KCL at the output junction. In solving op amp problems, KCL is applied at input terminals.

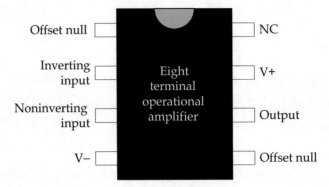

FIGURE 11.1: An eight-terminal operational amplifier. The terminal NC is not connected, and the two terminal offset nulls are used to correct imperfections (typically not connected). V^+ and V^- are terminal power to provide to the circuit. Keep in mind that a ground exists for both V^+ and V^-, a ground that is shared by other elements in the circuit. Modern operational amplifiers have ten or more terminals.

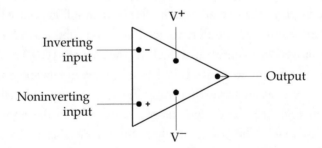

FIGURE 11.2: Circuit element symbol for the operational amplifier.

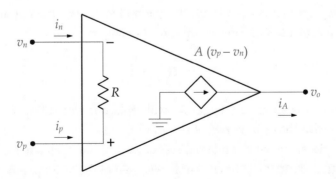

FIGURE 11.3: An internal model of the op amp. The internal resistance between the input terminals, R, is very large exceeding $1\,M\Omega$. The gain of the amplifier, A, is also large exceeding 10^4. Power supply terminals are omitted for simplicity.

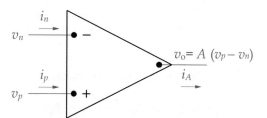

FIGURE 11.4: Idealized model of the op amp with the internal resistance, R, replaced by an open circuit.

Example 11.1. Find v_0 for the following circuit.

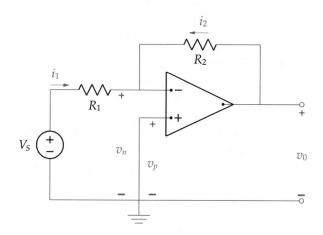

Solution. Using the op amp model of Fig. 11.4, we apply KCL at the inverting terminal giving

$$-i_1 - i_2 = 0$$

since no current flows into the op amp's input terminals. Replacing the current using Ohm's law gives

$$\frac{v_s - v_1}{R_1} + \frac{v_o - v_1}{R_2} = 0$$

Multiplying by $R_1 R_2$ and collecting like terms, we have

$$R_2 v_s = (R_1 + R_2) v_1 - R_1 v_o$$

Now $v_o = A(v_p - v_n)$ and since the noninverting terminal is connected to ground, $v_p = 0$ so

$$v_o = -A v_n$$

or

$$v_n = -\frac{v_o}{A}$$

Substituting v_n into the KCL inverting input equation gives,

$$R_s v_s = (R_1 + R_2)\left(-\frac{v_o}{A}\right) - R_1 v_o$$

$$= \left(\frac{R_1 + R_2}{A} + R_1\right) v_o$$

or

$$v_o = \frac{-R_2 v_s}{\left(R_1 + \dfrac{R_1 + R_2}{A}\right)}$$

As A goes to infinity, the previous equation goes to

$$v_o = -\frac{R_2}{R_1} v_s$$

Interestingly, with A going to infinity, v_0 remains finite due to the resistor R_2. This happens because a negative feedback path exists between the output and the inverting input terminal through R_2. This circuit is called an inverting amplifier with an overall gain of $-\frac{R_2}{R_1}$.   ■

An operational amplifier with a gain of infinity is known as an ideal op amp. Because of the infinite gain, there must be a feedback path between the output and input and we cannot connect a voltage source directly between the inverting and noninverting input terminals. When analyzing an ideal op amp circuit, we simplify the analysis by letting

$$v_n = v_p$$

Consider the previous example. With $v_p = 0$ means $v_n = 0$. Applying KCL at the inverting input gives

$$-\frac{v_s}{R_1} + \frac{-v_o}{R_2} = 0$$

or

$$v_o = -\frac{R_2}{R_1} v_s$$

Notice how simple the analysis becomes when we assume $v_n = v_p$. Keep in mind that this approximation is valid as long as A is very large (infinity) and a feedback is included.

Example 11.2. Find the overall gain for the following circuit.

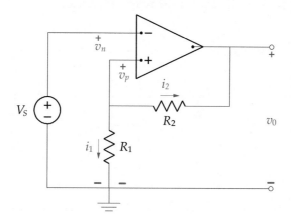

Solution. Assuming the op amp is ideal, we start with $v_n = v_p$. Then since the op amp's noninverting terminal is connected to the source, $v_n = v_p = v_s$. Because no current flows into the op amp, by KCL we have

$$i_1 + i_2 = 0$$

and

$$\frac{v_s}{R_1} + \frac{v_s - v_o}{R_2} = 0$$

or

$$v_o = \left(\frac{R_1 + R_2}{R_1}\right) v_s$$

The overall gain is

$$\frac{v_o}{v_s} = \frac{R_1 + R_2}{R_1}$$

This circuit is a noninverting op amp circuit used to amplify the source input. Amplifiers are used in most all clinical instrumentation from ECK, EEG, EOG, etc. ∎

The next example describes a summing op amp circuit.

Example 11.3. Find the overall gain for the following circuit.

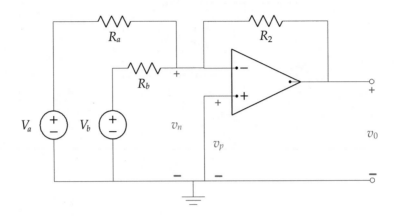

Solution. As before we start the solution with $v_n = v_p$ and note that the noninverting input is connected to ground, yielding $v_n = v_p = 0\,\mathrm{V}$. Applying KCL at the inverting input node gives

$$-\frac{V_a}{R_a} - \frac{V_b}{R_b} - \frac{v_o}{R_2} = 0$$

or

$$v_o = -\left(\frac{R_2}{R_a}V_a + \frac{R_2}{R_b}V_b\right)$$

We can add additional source resistor inputs, so that in general

$$v_o = -\left(\frac{R_2}{R_a}V_a + \frac{R_2}{R_b}V_b + \cdots + \frac{R_2}{R_m}V_m\right)$$ ■

Our next op amp circuit provides an output proportional to the difference of two input voltages. This op amp is often referred to as a differential amplifier.

Example 11.4. Find the overall gain for the following circuit.

Solution. Assuming an ideal op amp, we note no current flows into the input terminals and that $v_n = v_p$. Apply KCL at the inverting input terminal gives

$$\frac{v_n - V_a}{R_1} + \frac{v_n - v_o}{R_2} = 0$$

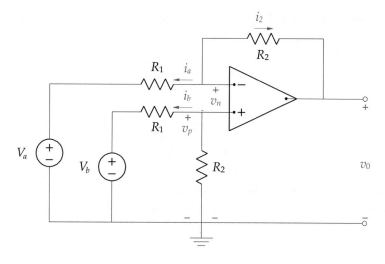

and

$$(R_1 + R_2)v_n - R_2 V_a = R_1 v_o$$

The previous equation involves two unknowns, thus we need another equation easily found by applying voltage divider at the noninverting input.

$$v_p = \frac{R_2}{R_1 + R_2} v_b = v_n$$

Substituting this result for v_n into the KCL equation at the inverting terminal gives

$$R_2 V_b - R_2 V_a = R_1 v_o$$

or

$$v_o = \frac{R_2}{R_1}(V_b - V_a)$$

As shown, this op amp circuit, also known as the differential amplifier, subtracts the weighted input signals. This amplifier is used for bipolar measurements involving ECG and EEG as the typical recording is obtained between two bipolar input terminals. Ideally, the measurement contains only the signal of interest uncontaminated by noise from the environment. The noise is typically called *common-mode signal.* Common-mode signal comes from lighting, 60-Hz power line signals, inadequate grounding and power supply leakage. A differential amplifier with appropriate filtering can reduce the impact of the common-mode signal. ∎

The response of a differential amplifier can be decomposed into differential-mode and common-mode components,

$$v_{dm} = v_b - v_a$$

and

$$v_{cm} = \frac{(v_a + v_b)}{2}$$

As described, the common-mode signal is the average of the input voltages. Using the two previous equations, one can solve v_a and v_b in terms of v_{dm} and v_{cm} as

$$v_a = v_{cm} - \frac{v_{dm}}{2}$$

and

$$v_b = v_{cm} + \frac{v_{dm}}{2}$$

When substituted into the response in Exercise 44 gives

$$v_0 = \left(\frac{R_1 R_2 - R_1 R_2}{R_1(R_1 + R_2)} \right) v_{cm} + \left(\frac{R_2(R_1 + R_2) + R_2(R_1 + R_2)}{2R_1(R_1 + R_2)} \right) v_{dm} = A_{cm} v_{cm} + A_{dm} v_{dm}$$

Notice the term multiplying, v_{cm}, A_{cm}, is zero, characteristic of the ideal op amp that amplifies only the differential-mode of the signal. Since real amplifiers are not ideal and resistors are not truly exact, the common-mode gain is not zero. So when one designs a differential amplifier, the goal is to keep A_{cm} as small as possible and A_{dm} as large as possible.

The rejection of the common-mode signal is called *common-mode rejection*, and the measure of how ideal the differential amplifier is called the *common-mode rejection ratio*, given as

$$CMRR = 20 \log_{10} \left| \frac{A_{dm}}{A_{cm}} \right|$$

where the larger the value of *CMRR* the better. Values of *CMRR* for a differential amplifier for EEG, ECG, and EMG is 100–120 db.

The general approach to solving op amp circuits is to first assume that the op amp is ideal and $v_p = v_n$. Next, we apply KCL or KVL at the two input terminals. In more complex circuits, we continue to apply our circuit analysis tools to solve the problem as the next example illustrates.

Example 11.5. Find v_0 for the following circuit.

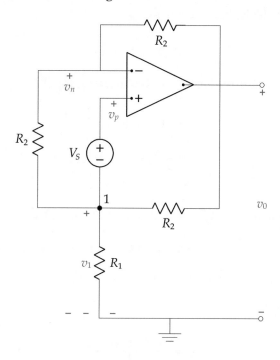

Solution. With $v_n = v_p$, we apply KCL at the inverting input

$$\frac{v_n - v_1}{R_2} + \frac{v_n - v_o}{R_2} = 0$$

and

$$2v_n - v_1 - v_o = 0$$

Next, we apply KVL from ground to node 1 to the noninverting input and back to ground giving

$$-v_1 - V_s + v_p = 0$$

and with $v_n = v_p$ we have $v_n - v_1 = V_s$.

Now, we apply KCL at node 1, noting no current flows into the noninverting input terminal

$$\frac{v_1}{R_1} + \frac{v_1 - v_o}{R_2} + \frac{v_1 - v_n}{R_2} = 0$$

Combining like terms in the previous equation gives

$$-R_1 v_n + (2R_1 + R_2)v_1 - R_1 v_o = 0$$

With three equations and three unknowns, we first eliminate v_1 by subtracting the inverting input KCL equation by the KVL equation giving

$$v_1 = v_o - 2V_s$$

Next, we eliminate v_n by substituting v_1 into the inverting input KCL equation as follows

$$
\begin{aligned}
v_n &= \frac{1}{2}(v_1 + v_o) \\
&= \frac{1}{2}(v_o - 2V_s + v_o) \\
&= v_o - V_s
\end{aligned}
$$

Finally, we substitute the solutions for v_1 and v_n into the node 1 KCL equation giving

$$-R_1 v_n + (2R_1 + R_2)v_1 - R_1 v_o = 0$$

$$-R_1(v_o - V_s) + (2R_1 + R_2)(v_o - 2V_s) - R_1 v_o = 0$$

After simplification, we have

$$v_o = \frac{(3R_1 + 2R_2)}{R_2} V_s$$

∎

The next two examples illustrate an op amp circuit that differentiates and integrates by using a capacitor.

Example 11.6. Find v_0 for the following circuit.

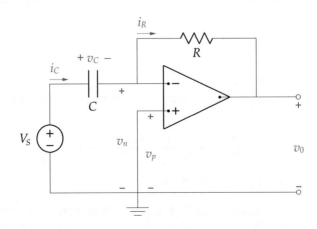

Solution. With the noninverting input connected to ground, we have $v_p = 0 = v_n$. From KVL

$$v_C = V_s$$

and it follows that

$$i_C = C\frac{dv_C}{dt} = C\frac{dV_s}{dt}$$

Since no current flows into the op amp, $i_C = i_R$. With

$$i_R = \frac{v_n - v_o}{R} = -\frac{v_o}{R}$$

and

$$i_C = C\frac{dV_s}{dt} = i_R = -\frac{v_o}{R}$$

we have

$$v_o = -RC\frac{dV_s}{dt}$$

If $R = \frac{1}{C}$, the circuit in this example differentiates the input, $v_o = -\frac{dV_s}{dt}$. ∎

Example 11.7. Find v_0 for the following circuit.

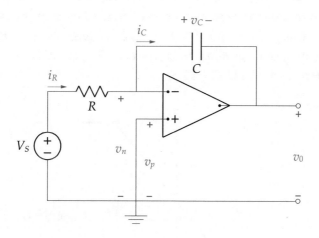

Solution. It follows that

$$v_n = v_p = 0$$

and

$$i_C = i_R = \frac{V_s}{R}$$

Therefore

$$v_C = \frac{1}{C} \int_{-\infty}^{t} i_C d\lambda = \frac{1}{C} \int_{-\infty}^{t} \frac{V_s}{R} d\lambda$$

From KVL, we have

$$v_C + v_o = 0$$

and

$$v_o = -\frac{1}{RC} \int_{-\infty}^{t} V_s d\lambda$$

With $R = \frac{1}{C}$, the circuit operates as an integrator

$$v_o = -\int_{-\infty}^{t} V_s d\lambda$$

∎

11.1 VOLTAGE CHARACTERISTICS OF THE OP AMP

In the past examples involving the op amp, we have neglected to consider the supply voltage (shown in Fig. 11.2) and that the output voltage of an ideal op amp is constrained to operate between the supply voltages V^+ and V^-. If analysis determines v_0 is greater than V^+, v_0 saturates at V^+. If analysis determines v_0 is less than V^-, v_0 saturates at V^-. The output voltage characteristics are shown in Fig. 11.5.

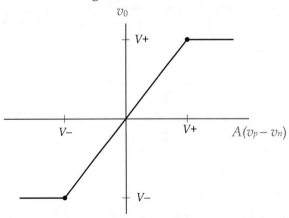

FIGURE 11.5: Voltage characteristics of an op amp.

Example 11.8. For the circuit shown in Ex. 11.5, let $V^+ = +10\,\text{V}$ and $V^- = -10\,\text{V}$. Graph the output voltage characteristics of the circuit.

Solution. The solution for Ex. 11.5 is

$$v_o = \left(\frac{3R_1 + 2R_2}{R_2} \right) V_s$$

which saturates whenever v_0 is less than V^- and greater than V^+ as shown in the following graph.

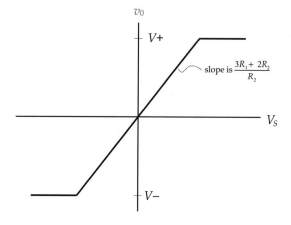

CHAPTER 12

Time-Varying Signals

An alternating current (AC) or sinusoidal source of 50 or 60 Hz is common throughout the world as a power source supplying energy for most equipment and other devices. While most of this chapter has focused on the transient response; when dealing with sinusoidal sources, attention is now focused on the steady-state or forced response. In bioinstrumentation, analysis in the steady state simplifies the design by focusing only on the steady-state response, which is where the device actually operates. A sinusoidal voltage source is a time-varying signal given by

$$v_s = V_m \cos(\omega t + \phi) \qquad (12.1)$$

where the voltage is defined by angular frequency (ω in radians/s), phase angle (ϕ in radians or degrees), and peak magnitude (V_m). The period of the sinusoid T is related to frequency f (Hz or cycles/s) and angular frequency by

$$\omega = 2\Pi f = \frac{2\Pi}{T} \qquad (12.2)$$

An important metric of a sinusoid is its *rms value* (square root of the mean value of the squared function), given by

$$V_{rms} = \sqrt{\frac{1}{T} \int_0^T V_m^2 \cos^2(\omega t + \phi) dt} \qquad (12.3)$$

which reduces to $V_{rms} = \frac{V_m}{\sqrt{2}}$.

To appreciate the response to a time-varying input, $v_s = V_m \cos(\omega t + \phi)$, consider the circuit shown in Fig. 12.1 in which the switch is closed at $t = 0$ and there is no initial energy stored in the inductor.

Applying KVL to the circuit gives

$$L\frac{di}{dt} + iR = V_m \cos(\omega t + \phi)$$

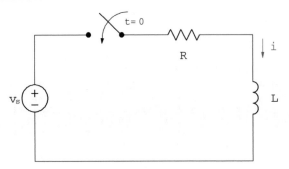

FIGURE 12.1: A RL circuit with sinusoidal input.

and after some work, the solution is

$$i = i_n + i_f$$

$$= \frac{-V_m}{\sqrt{R^2 + \omega^2 L^2}} \cos\left(\phi - \frac{\omega L}{R}\right) e^{-\frac{R}{L}t} + \frac{V_m}{\sqrt{R^2 + \omega^2 L^2}} \cos\left(\omega t + \phi - \frac{\omega L}{R}\right)$$

The first term is the natural response that goes to zero as t goes to infinity. The second term is the forced response that has the same form as the input (i.e., a sinusoid with the same frequency ω, but a different phase angle and maximum amplitude). If all you are interested in is the steady-state response as in most bioinstrumentation applications, then the only unknowns are the response amplitude and phase angle. The remainder of this section deals with techniques involving the *phasor* to efficiently find these unknowns.

12.1   PHASORS

The phasor is a complex number that contains amplitude and phase angle information of a sinusoid, and for the signal in Eq. (12.1) is expressed as

$$\mathbf{V} = V_m e^{j\phi} = V_m \underline{/\phi} \qquad (12.4)$$

In Eq. (12.1), by practice, the angle in the exponential is written in radians, and in the $\underline{/\phi}$ notation, in degrees. Work in the phasor domain involves the use of complex algebra in moving between the time and phasor domain, therefore, the rectangular form of the phasor is also used, given as

$$\mathbf{V} = V_m (\cos\phi + j\sin\phi) \qquad (12.5)$$

12.2 PASSIVE CIRCUIT ELEMENTS IN THE PHASOR DOMAIN

To use phasors with passive circuit elements for steady-state solutions, the relationship between voltage and current is needed for the resistor, inductor and capacitor. Assume that

$$i = I_m \cos(\omega t + \theta)$$

$$I = I_m\underline{|\theta} = I_m e^{j\theta}$$

For a resistor,

$$v = IR = RI_m \cos(\omega t + \theta)$$

and the phasor of v is

$$\mathbf{V} = RI_m\underline{|\theta} = R\mathbf{I} \qquad (12.6)$$

Note that there is no phase shift for the relationship between the phasor current and voltage for a resistor.

For an inductor,

$$v = L\frac{di}{dt} = -\omega L I_m \sin(\omega t + \theta) = -\omega L I_m \cos(\omega t + \theta - 90°)$$

and the phasor of v is

$$\begin{aligned}
\mathbf{V} &= -\omega L I_m\underline{|\theta - 90°} = -\omega L I_m e^{j(\theta - 90°)} \\
&= -\omega L I_m e^{j\theta} e^{-j90°} = -\omega L I_m e^{j\theta}(-j) \\
&= j\omega L I_m e^{j\theta} \\
&= j\omega L\mathbf{I}
\end{aligned} \qquad (12.7)$$

Note that inductor current and voltage are out of phase by 90°, that is current lags behind voltage by 90°.

For a capacitor, define $v = V_m \cos(\omega t + \theta)$ and $\mathbf{V} = V_m\underline{|\theta}$. Now

$$i = C\frac{dv}{dt} = C\frac{d}{dt}(V_m \cos(\omega t + \theta))$$

$$= -CV_m\omega \sin(\omega t + \theta) = -CV_m\omega \cos(\omega t + \theta - 90°)$$

And the phasor for i is

$$\begin{aligned}
\mathbf{I} &= -\omega C V_m\underline{|\theta - 90°} = -\omega C V_m e^{j\theta} e^{-j90°} \\
&= -\omega C V_m e^{j\theta} (\cos(90°) - j\sin(90°)) \\
&= j\omega C V_m e^{j\theta} \\
&= j\omega C\mathbf{V}
\end{aligned}$$

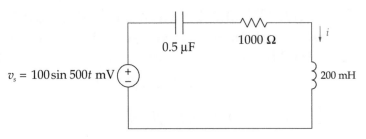

FIGURE 12.2: A circuit diagram.

or

$$\mathbf{V} = \frac{1}{j\omega C}\mathbf{I} = \frac{-j}{\omega C}\mathbf{I} \tag{12.8}$$

Note that capacitor current and voltage are out of phase by $90°$, that is voltage lags behind current by $90°$.

Equations (12.6)–(12.8) all have the form of $\mathbf{V} = Z\mathbf{I}$, where Z represents the impedance of the circuit element and is, in general, a complex number, with units of Ohms. The impedance for the resistor is R, the inductor, $j\omega L$, and the capacitor, $\frac{-j}{\omega C}$. The impedance is a complex number and not a phasor even though it may look like one. The imaginary part of the impendence is called reactance.

The final part to working in the phasor domain is to transform a circuit diagram from the time to phasor domain. For example, the circuit shown in Fig. 12.2 is transformed into the phasor domain, shown in Fig. 12.3, by replacing each circuit element with their impedance equivalent and sources by their phasor. For the voltage source, we have

$$v_s = 100 \sin 500t = 100 \cos(500t - 90°) \text{ mV} \leftrightarrow 500 \underline{|-90°} \text{ mV}$$

For the capacitor, we have

$$0.5\mu F \leftrightarrow \frac{-j}{\omega C} = -j4000\,\Omega$$

FIGURE 12.3: Phasor and impedance equivalent circuit for Fig. 12.2.

For the resistor, we have

$$1000\,\Omega \leftrightarrow 1000\,\Omega$$

For the inductor, we have

$$200\,\text{mH} \leftrightarrow j\omega L = j100\,\Omega$$

Each of the elements are replaced by their phasor and impedance equivalents as shown in Fig. 12.3.

12.3 KIRCHHOFF'S LAWS AND OTHER TECHNIQUES IN THE PHASOR DOMAIN

It is fortunate that all of the material presented before in this chapter involving Kirchhoff's current and voltage laws, and all the other techniques apply to phasors. That is, for KVL, the sum of phasor voltages around any closed path is zero

$$\sum \mathbf{V}_i = 0 \qquad\qquad (12.9)$$

and for KCL, the sum of phasor currents leaving any node is zero

$$\sum \mathbf{I} = 0 \qquad\qquad (12.10)$$

Impedances in series are given by

$$Z = Z_1 + \cdots + Z_n \qquad\qquad (12.11)$$

Impedances in parallel are given by

$$Z = \frac{1}{\frac{1}{Z_1} + \cdots + \frac{1}{Z_n}} \qquad\qquad (12.12)$$

The node-voltage method, as well as superposition, Thévenin equivalent circuits is applicable in the phasor domain. The following two examples illustrate the process, with the most difficult aspect involving complex algebra.

Example Problem 12.1 For the circuit shown in Fig. 12.3, find the steady-state response i.

Solution. The impedance for the circuit is

$$Z = -j4000 + 1000 + j100 = 1000 - j3900\,\Omega$$

Using Ohm's law,

$$\mathbf{I} = \frac{\mathbf{V}}{Z} = \frac{0.5\,\lfloor -90^\circ}{1000 - j3900} = \frac{0.5\,\lfloor -90^\circ}{4026\lfloor -76^\circ} = 124\lfloor -14^\circ\,\mu\text{A}$$

Returning to the time domain, the steady-state current is

$$i = 124 \cos{(500t - 14°)} \, \mu A$$

■

Example Problem 12.2. Find the steady-state response v using the node-voltage method for the following circuit.

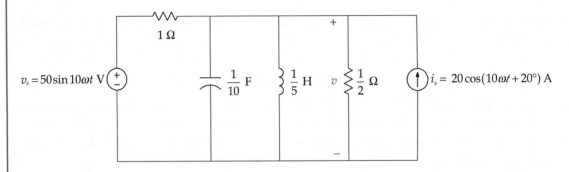

Solution. The first step is to transform the circuit elements into their impedances, which for the capacitor and inductor are

$$\frac{1}{10} F \leftrightarrow \frac{-j}{\omega C} = -j \, \Omega$$

$$\frac{1}{5} H \leftrightarrow j\omega L = j2 \, \Omega$$

The phasors for the two sources are:

$$v_s = 50 \sin \omega t \, V \leftrightarrow \mathbf{V_s} = 50 \underline{|-90°} V$$
$$i_s = 20 \cos{(\omega t + 20°)} \, A \leftrightarrow \mathbf{I_s} = 20 \underline{|20°}$$

Since the two resistors retain their values, the phasor drawing of the circuit is shown in the following figure with the ground at the lower node.

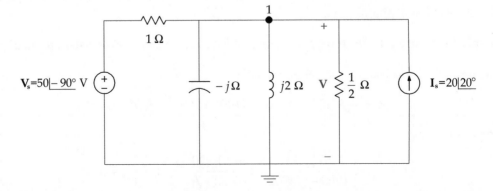

Writing the node-voltage equation for node 1 gives

$$\mathbf{V} - 50\underline{|-90°} + \frac{\mathbf{V}}{-j} + \frac{\mathbf{V}}{j2} + 2\mathbf{V} - 20\underline{|20°} = 0$$

Collecting like terms, converting to rectangular form and converting to polar form gives

$$\mathbf{V}\left(3 + \frac{j}{2}\right) = 50\underline{|-90°} + 20\underline{|20°}$$

$$\mathbf{V}\left(3 + \frac{j}{2}\right) = -50j + 18.8 + j6.8 = 18.8 - j43.2$$

$$\mathbf{V} \times 3.04\underline{|9.5°} = 47.1\underline{|-66.5°}$$

$$\mathbf{V} = \frac{47.1\underline{|-66.5°}}{3.04\underline{|9.5°}} = 15.5\underline{|-76°}$$

The steady-state solution is

$$v = 15.6\cos\left(10t - 76°\right) \text{ V} \qquad \blacksquare$$

CHAPTER 13

Active Analog Filters

This section presents several active analog filters involving the op amp. Passive analog filters use passive circuit elements: resistors, capacitors and inductors. To improve performance in a passive analog filter, the resistive load at the output of the filter is usually increased. By using the op amp, fine control of the performance is achieved without increasing the load at the output of the filter. Filters are used to modify the measured signal by removing noise. A filter is designed in the frequency domain so that the measured signal to be retained is passed through and noise is rejected.

Shown in Fig. 13.1 are the frequency characteristics of four filters: low-pass, high-pass, band-pass and notch filters. The signal that is passed through the filter is indicated by the frequency interval called the passband. The signal that is removed by the filter is indicated by the frequency interval called the stopband. The magnitude of the filter, $|H(j\omega)|$, is one in the passband and zero in the stopband. The low-pass filter allows slowly changing signals with frequency less than ω_1 pass through the filter, and eliminates any signal or noise above ω_1. The high-pass filter allows quickly changing signals with frequency greater than ω_2 to pass through the filter, and eliminates any signal or noise with frequency less than ω_2. The band-pass filter allows signals in the frequency band greater than ω_1 and less than ω_2 to pass through the filter, and eliminates any signal or noise outside this interval. The notch filter allows signals in the frequency band less than ω_1 and greater than ω_2 to pass through the filter, and eliminates any signal or noise outside this interval. The frequencies ω_1 and ω_2 are typically called cutoff frequencies for the low-pass and high-pass filters.

In reality, any real filter cannot possibly have these ideal characteristics, but instead has a smooth transition from the passband to the stopband, as shown, for example in Fig. 13.2; the reason for this behavior is described in a later chapter. Further, it is sometimes convenient to include both amplification and filtering in the same circuit, so the maximum of the magnitude does not need to be one, but can be a value of M specified by the needs of the application.

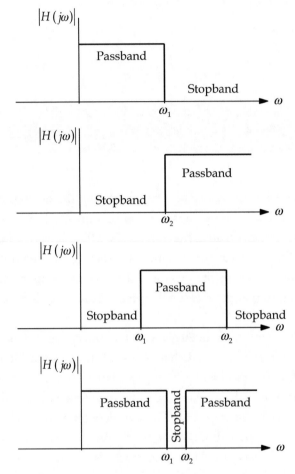

FIGURE 13.1: Ideal magnitude–frequency response for four filters, from top to bottom: low-pass, high-pass, band-pass and notch.

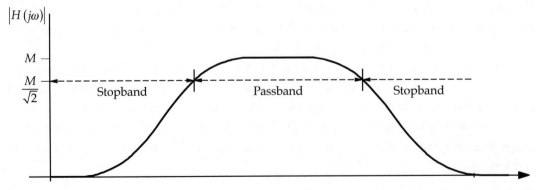

FIGURE 13.2: A realistic magnitude–frequency response for a band-pass filter. Note that the magnitude M does not necessarily need to be one. The passband is defined as the frequency interval when the magnitude is greater than $\frac{M}{\sqrt{2}}$.

To determine the filter's performance, the filter is driven by a sinusoidal input. One varies the input over the entire spectrum of interest (at discrete frequencies) and records the output magnitude. The critical frequencies are when $|H(j\omega)| = \frac{M}{\sqrt{2}}$.

Example 13.1. Using the low-pass filter in the following circuit, design the filter to have a gain of 5 and a cutoff frequency of 500 rad/s.

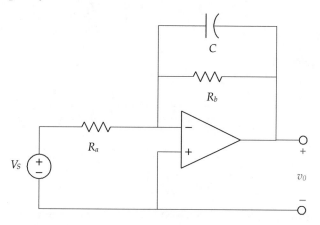

Solution. By treating the op amp as ideal, note that the noninverting input is connected to ground, and therefore, the inverting input is also connected to ground. The operation of this filter is readily apparent for at low frequencies, the capacitor acts like an open circuit, reducing the circuit to an inverting amplifier that passes low frequency signals. At high frequencies, the capacitor acts like a short circuit, which connects the output terminal to the inverting input and ground.

The phasor method will be used to solve this problem by first transforming the circuit into the phasor domain as shown in the following figure.

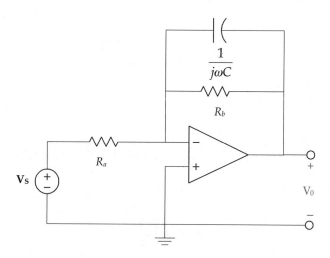

Summing the currents leaving the inverting input gives

$$-\frac{V_s}{R_a} - \frac{V_0}{\frac{1}{j\omega C}} - \frac{V_0}{R_b} = 0$$

Collecting like terms and rearranging yields

$$-V_0 \left(\frac{1}{\frac{1}{j\omega C}} + \frac{1}{R_b} \right) = \frac{V_s}{R_a}$$

After further manipulation,

$$\frac{V_0}{V_s} = -\frac{1}{R_a} \left(\frac{1}{\frac{1}{j\omega C} + \frac{1}{R_b}} \right) = -\frac{1}{R_a} \left(\frac{1}{j\omega C + \frac{1}{R_b}} \right)$$

$$\frac{V_0}{V_s} = -\frac{1}{R_a C} \left(\frac{1}{j\omega + \frac{1}{R_b C}} \right)$$

Similar to the reasoning for the characteristic equation for a differential equation, the cutoff frequency is defined as $\omega_c = \frac{1}{R_b C}$ (i.e., the denominator term, $j\omega + \frac{1}{R_b C}$ set equal to zero). Thus, with the cutoff frequency set at $\omega_c = 500$ rad/s, then $\frac{1}{R_b C} = 500$. The cutoff frequency is also defined as when $|H(j\omega)| = \frac{M}{\sqrt{2}}$, where $M = 5$. The magnitude of $\frac{V_0}{V_s}$ is given by

$$\left| \frac{V_0}{V_s} \right| = \frac{\frac{1}{R_a C}}{\sqrt{\omega^2 + \left(\frac{1}{R_b C} \right)^2}}$$

and at the cutoff frequency, $\omega_c = 500$ rad/s,

$$\frac{5}{\sqrt{2}} = \frac{\frac{1}{R_a C}}{\sqrt{\omega_c^2 + \left(\frac{1}{R_b C} \right)^2}}$$

With $\frac{1}{R_b C} = 500$, the magnitude is

$$\frac{5}{\sqrt{2}} = \frac{\frac{1}{R_a C}}{\sqrt{\omega_c^2 + \left(\frac{1}{R_b C} \right)^2}} = \frac{\frac{1}{R_a C}}{\sqrt{500^2 + 500^2}} = \frac{\frac{1}{R_a C}}{500\sqrt{2}}$$

which gives

$$R_a C = \frac{1}{2500}$$

Since we have three unknowns and two equations ($R_a C = \frac{1}{2500}$ and $\frac{1}{R_b C} = 500$), there are an infinite number of solutions. Therefore, one can select a convenient value for one of the elements, say $R_a = 20\,k\Omega$, and the other two elements are determined as

$$C = \frac{1}{2500 \times R_a} = \frac{1}{2500 \times 20000} = 20\,nF$$

and

$$R_b = \frac{1}{500 \times C} = \frac{1}{500 \times 20 \times 10^{-9}} = 100\,k\Omega$$

A plot of the magnitude versus frequency is shown in the following figure. As can be seen, the cutoff frequency gives a value of magnitude equal to 3.53 at 100 Hz, which is the design goal.

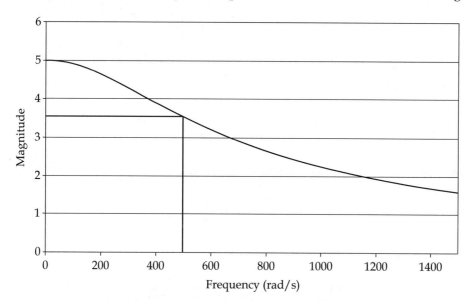

Example 13.2. Using the high-pass filter in the following circuit, design the filter to have a gain of 5 and a cutoff frequency of 100 rad/s.

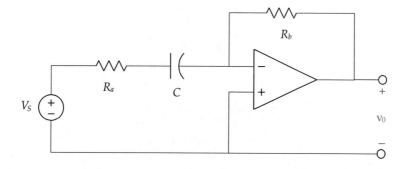

Solution. Since the op amp is assumed ideal, and the noninverting input is connected to ground, therefore the inverting input is also connected to ground. The operation of this filter is readily apparent for at low frequencies, the capacitor acts like an open circuit and so no input voltage is seen at the noninverting input. Since there is no input, then the output is zero. At high frequencies, the capacitor acts like a short circuit, which reduces the circuit to an inverting amplifier that passes through high frequency signals.

As before, the phasor method will be used to solve this problem by first transforming the circuit into the phasor domain as shown in the following figure.

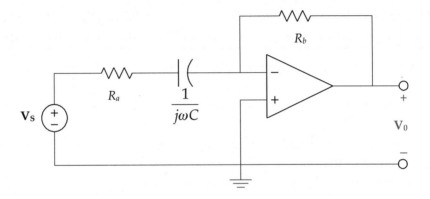

Summing the currents leaving the inverting input gives

$$-\frac{V_s}{R_a + \frac{1}{j\omega C}} - \frac{V_0}{R_b} = 0$$

Rearranging yields

$$\frac{V_0}{V_s} = -\frac{R_b}{R_a + \frac{1}{j\omega C}} = -\frac{R_b}{R_a} \frac{j\omega}{j\omega + \frac{1}{R_a C}}$$

At cutoff frequency $\omega_c = 100 \text{ rad/s} = \frac{1}{R_a C}$. The magnitude of $\frac{V_0}{V_s}$ is given by

$$\left|\frac{V_0}{V_s}\right| = \frac{R_b}{R_a} \frac{\omega}{\sqrt{\omega^2 + \left(\frac{1}{R_a C}\right)^2}}$$

and at the cutoff frequency,

$$\frac{5}{\sqrt{2}} = \frac{R_b}{R_a} \frac{\omega_c}{\sqrt{\omega_c^2 + \left(\frac{1}{R_a C}\right)^2}}$$

With $\frac{1}{R_aC} = 100$ and $\omega_c = 100$ rad/s, gives

$$\frac{5}{\sqrt{2}} = \frac{R_b}{R_a} \frac{\omega_c}{\sqrt{\omega_c^2 + \left(\frac{1}{R_aC}\right)^2}} = \frac{R_b}{R_a} \frac{\frac{1}{R_aC}}{\sqrt{100^2 + 100^2}} = \frac{R_b}{R_a} \frac{100}{100} = \frac{R_b}{\sqrt{2}R_a}$$

Thus $\frac{R_b}{R_a} = 5$. Since we have three unknowns and two equations, one can select a convenient value for one of the elements, say $R_b = 20\,\text{k}\Omega$, and the other two elements are determined as

$$R_a = \frac{R_b}{5} = \frac{20000}{5} = 4\,\text{k}\Omega$$

and

$$C = \frac{1}{100\,R_a} = \frac{1}{100 \times 4000} = 2.5\,\mu\text{F}$$

A plot of the magnitude versus frequency is shown in the following figure. As can be seen, the cutoff frequency gives a value of magnitude equal to 3.53 at 500 Hz, which is the design goal.

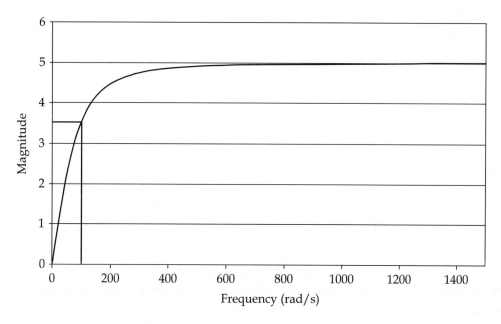

The next example demonstrates the technique to create a band-pass filters (which require two cutoff frequencies).

Example 13.3. Using the band-pass filter in the following circuit, design the filter to have a gain of 5 and pass through frequencies from 100 to 500 rad/s.

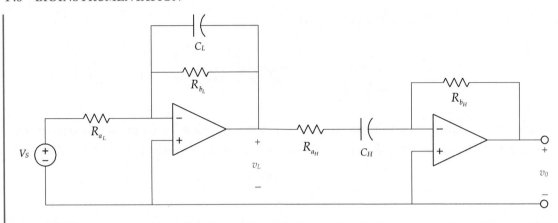

Solution. As usual, the design of the filter is done in the phasor domain, and makes use of work done in the previous two examples. Note the elements around the op amp on the left are the low-pass filter circuit elements, and those on the right, the high-pass filter. In fact, when working with op amps, filters can be cascaded together to form other filters; thus a low-pass and high-pass filter cascaded together form a band-pass. The phasor domain circuit is given in the next figure.

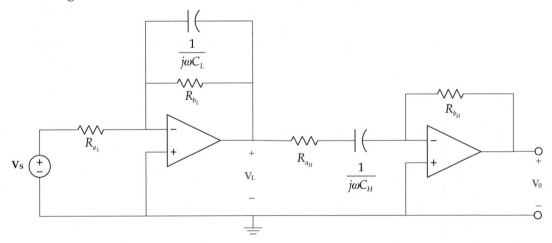

As before, the noninverting input to the op amps are connected to ground, which means that the inverting input is also connected to ground. Summing the currents leaving the inverting input for each op amp gives

$$-\frac{V_s}{R_{a_L}} - \frac{V_L}{\frac{1}{j\omega C_L}} - \frac{V_L}{R_{b_L}} = 0$$

$$-\frac{V_L}{R_{a_H} + \frac{1}{j\omega C_H}} - \frac{V_0}{R_{b_H}} = 0$$

Solving the first equation for $\mathbf{V_L}$ gives

$$\mathbf{V_L} = -\frac{1}{R_{a_L} C_L} \left(\frac{1}{j\omega + \frac{1}{R_{b_L} C_L}} \right) \mathbf{V_s}$$

Solving the second equation for $\mathbf{V_0}$ gives

$$\mathbf{V_0} = -\frac{R_{b_H}}{R_{a_H}} \frac{j\omega}{j\omega + \frac{1}{R_{a_H} C_H}} \mathbf{V_L}$$

Substituting $\mathbf{V_L}$ into the previous equation yields

$$\mathbf{V_0} = \frac{R_{b_H}}{R_{a_H}} \frac{j\omega}{j\omega + \frac{1}{R_{a_H} C_H}} \times \frac{1}{R_{a_L} C_L} \left(\frac{1}{j\omega + \frac{1}{R_{b_L} C_L}} \right) \mathbf{V_s}$$

The form of the solution is simply the product of each filter. The magnitude of the filter is

$$\left| \frac{\mathbf{V_0}}{\mathbf{V_s}} \right| = \frac{R_{b_H}}{R_{a_H}} \frac{\omega}{\sqrt{\omega^2 + \left(\frac{1}{R_{a_H} C_H} \right)^2}} \frac{\frac{1}{R_{a_L} C_L}}{\sqrt{\omega^2 + \left(\frac{1}{R_{b_L} C_L} \right)^2}}$$

Since there are two cutoff frequencies, two equations evolve,

$$\omega_{c_H} = \frac{1}{R_{a_H} C_H} = 100 \text{ rad/s}$$

and

$$\omega_{c_L} = \frac{1}{R_{b_L} C_L} = 500 \text{ rad/s}$$

At the either cutoff frequency, the magnitude is $\frac{5}{\sqrt{2}}$, such that at $\omega_{c_H} = 100$ rad/s

$$\frac{5}{\sqrt{2}} = \frac{R_{b_H}}{R_{a_H}} \frac{\omega_{c_H}}{\sqrt{\omega_{c_H}^2 + \left(\frac{1}{R_{a_H} C_H} \right)^2}} \frac{\frac{1}{R_{a_L} C_L}}{\sqrt{\omega_{C_H}^2 + \left(\frac{1}{R_{b_L} C_L} \right)^2}}$$

$$= \frac{R_{b_H}}{R_{a_H}} \frac{100}{\sqrt{100^2 + 100^2}} \frac{\frac{1}{R_{a_L} C_L}}{\sqrt{100^2 + 500^2}}$$

Therefore,

$$500\sqrt{26} = \frac{R_{b_H}}{R_{a_H} R_{a_L} C_L}$$

The other cutoff frequency gives the same result as the previous equation. There are now three equations ($\frac{1}{R_{a_H} C_H} = 100$, $\frac{1}{R_{b_L} C_L} = 500$ and $500\sqrt{26} = \frac{R_{b_H}}{R_{a_H} R_{a_L} C_L}$), and six unknowns. For convenience, set $R_{b_L} = 100\,\text{k}\Omega$ and $R_{a_H} = 100\,\text{k}\Omega$, which gives $C_L = \frac{1}{500 R_{b_L}} = 20\text{nF}$ and $C_H = \frac{1}{100 R_{a_H}} = 0.1\,\mu\text{F}$. Now from $500\sqrt{26} = \frac{R_{b_H}}{R_{a_H}} \frac{1}{R_{a_L} C_L}$,

$$\frac{R_{b_H}}{R_{a_L}} = 500\sqrt{26} C_L R_{a_H} = 5.099$$

Once again, one can specify one of the resistors, say $R_{a_L} = 10\,\text{k}\Omega$, giving $R_{b_H} = 50.099\,\text{k}\Omega$.

A plot of the magnitude versus frequency is shown in the following figure. As can be seen, the cutoff frequency gives a value of magnitude equal to 3.53 at 500 Hz, which is the design goal.

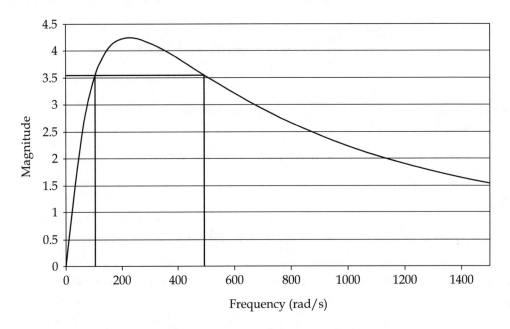

None of the filters in Examples 13.1–13.3 have the ideal characteristics of Fig. 13.1. To improve the performance from the passband to stopband in a low-pass filter with a sharper transition, one can cascade identical filters together, i.e., connect the output of the first filter to the input of the next filter and so on. The more cascaded filters, the better the performance. The magnitude of the overall filter is the product of the individual filter magnitudes.

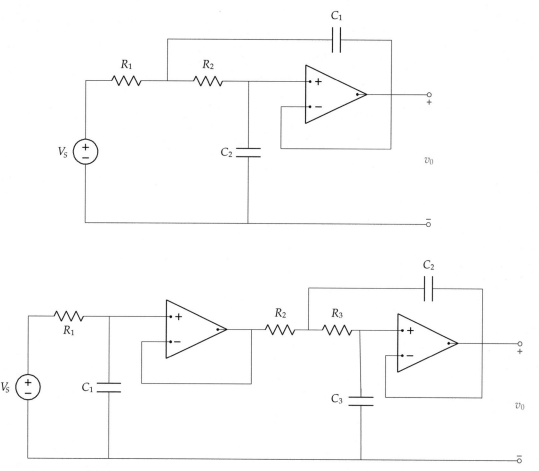

FIGURE 13.3: (Top) Second-order Butterworth low-pass filter. (Bottom) Third-order Butterworth low-pass filter.

While this approach is appealing for improving the performance of the filter, the overall magnitude of the filter does not remain a constant in the passband. Better filters are available with superior performance such as a Butterworth filter. Two Butterworth filters are shown in Fig. 13.3. Analysis of these filters is carried out in Exercises 183 and 184.

CHAPTER 14

Bioinstrumentation Design

Figure 14.1 described the various elements needed in a biomedical instrumentation system. The purpose of this type of instrument is to monitor the output of a sensor or sensors and to extract information from the signals that are produced by the sensors.

Acquiring a discrete-time signal and storing this signal in computer memory from a continuous-time signal is accomplished with an analog-to-digital (A/D) converter. The A/D converter uniformly samples the continuous-time waveform and transforms it into a sequence of numbers, one every t_k seconds. The A/D converter also transforms the continuous-time wave form into a digital signal (i.e., the amplitude takes one of 2^n discrete values) which are converted into computer words and stored in computer memory. To adequately capture the continuous-time signal, the sampling instants t_k must be selected carefully so that information is not lost. The minimum sampling rate is twice the highest frequency content of the signal (based on the sampling theorem from communication theory). Realistically, we often sample at five to ten times the highest frequency content of the signal so as to achieve better accuracy by reducing aliasing error.

14.1 NOISE

Measurement signals are always corrupted by noise in a biomedical instrumentation system. Interference noise occurs when unwanted signals are introduced into the system by outside sources, e.g. power lines and transmitted radio and television electromagnetic waves. This kind of noise is effectively reduced by careful attention to the circuit's wiring configuration to minimize coupling effects.

Interference noise is introduced by power lines (50 or 60 Hz), fluorescent lights, AM/FM radio broadcasts, computer clock oscillators, laboratory equipment, cellular phones, etc. Electromagnetic energy radiating from noise sources is injected into the amplifier circuit or into the patient by capacitive and/or inductive coupling. Even the action potentials from nerve conduction in the patient generate noise at the sensor/amplifier interface. Filters are used to reduce the noise and to maximize the signal-to-noise (S/N) ratio at the input of the A/D converter.

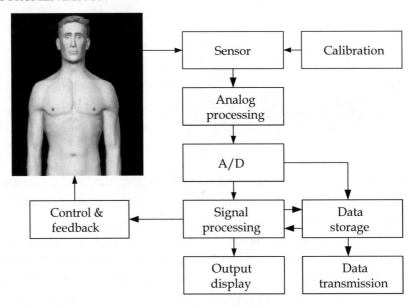

FIGURE 14.1: Basic instrumentation systems using sensors to measure a signal with data acquisition, storage and display capabilities, along with control and feedback.

Low frequency noise (amplifier DC offsets, sensor drift, temperature fluctuations, etc.) is eliminated by a high-pass filter with the cutoff frequency set above the noise frequencies and below the biological signal frequencies. High frequency noise (nerve conduction, radio broadcasts, computers, cellular phones, etc.) is reduced by a low-pass filter with the cutoff set below the noise frequencies and above the frequencies of the biological signal that is being monitored. Power-line noise is a very difficult problem in biological monitoring since the 50 or 60 Hz frequency is usually within the frequency range of the biological signal that is being measured. Band-stop filters are commonly used to reduce power-line noise. The notch frequency in these band-stop filters is set to the power-line frequency of 50 or 60 Hz with the cutoff frequencies located a few Hertz to either side.

The second type of corrupting signal is called inherent noise. Inherent noise arises from random processes that are fundamental to the operation of the circuit's elements and, hence, is reduced by good circuit design practice. While inherent noise can be reduced, it can never be eliminated. Low-pass filters can be used to reduce high frequency components. However, noise signals within the frequency range of the biosignal being amplified cannot be eliminated by this filtering approach.

14.2 COMPUTERS

Computers consist of three basic units: the central processing unit (CPU), the arithmetic and logic unit (ALU), and memory. The CPU directs the functioning of all other units and controls the flow of information among the units during processing procedures. It is controlled by program instructions. The ALU performs all arithmetic calculations (add, subtract, multiply, and divide) as well as logical operations (AND, OR, NOT) that compare one set of information to another.

Computer memory consists of read only memory (ROM) and random access memory (RAM). ROM is permanently programmed into the integrated circuit that forms the basis of the CPU and cannot be changed by the user. RAM stores information temporarily and can be changed by the user. RAM is where user-generated programs, input data, and processed data are stored.

Computers are binary devices that use the presence of an electrical signal to represent 1 and the absence of an electrical pulse to represent 0. The signals are combined in groups of 8 bits , a byte, to code information. A word is made up of 2 bytes. Most desktop computers that are available today are 32-bit systems, which means that they can address 4.29×10^9 locations in memory. The first microcomputers were 8-bit devices that could interact with only 256 memory locations.

Programming languages relate instructions and data to a fixed array of binary bits so that the specific arrangement has only one meaning. Letters of the alphabet and other symbols, e.g. punctuation marks, are represented by special codes. ASCII stands for the American Standard Code for Information Exchange. ASCII provides a common standard that allows different types of computers to exchange information. When word processing files are saved as text files, they are saved in ASCII format. Ordinarily, word processing files are saved in special program-specific binary formats, but almost all data analysis programs can import and export data in ASCII files.

The lowest level of computer languages is machine language and consists of the 0s and 1s that the computer interprets. Machine language represents the natural language of a particular computer. At the next level, assembly languages use English-like abbreviations for binary equivalents. Programs written in assembly language can manipulate memory locations directly. These programs run very quickly and are often used in data acquisition systems that must rapidly acquire a large number of samples, perhaps from an array of sensors, at a very high sampling rate.

Higher level languages, e.g. FORTRAN, PERL, and C++, contain statements that accomplish tasks that require many machine or assembly language statements. Instructions in these languages often resemble English and contain commonly used mathematical notations. Higher level languages are easier to learn than machine and assembly languages. Program instructions are designed to tell computers when and how to use various hardware components to solve specific problems. These instructions must be delivered to the CPU of a computer in the correct sequence in order to give the desired result.

When computers are used to acquire physiological data, programming instructions tell the computer when data acquisition should begin, how often samples should be taken from how many sensors, how long data acquisition should continue, and where the digitized data should be stored. The rate at which a system can acquire samples is dependent upon the speed of the computer's clock, e.g. 233 MHz, and the number of computer instructions that must completed in order to take a sample. Some computers can also control the gain on the input amplifiers so that signals can be adjusted during data acquisition. In other systems, the gain of the input amplifiers must be manually adjusted.

Exercises

1. Suppose the current flowing through the circuit element in Fig. 3.5 is

$$i(t) = \begin{cases} 0 & t < 0 \\ 5e^{-2t} \text{ A} & t \geq 0 \end{cases}$$

Find $q(t)$.

2. The charge entering the upper terminal in the circuit element in Fig. 3.5 is $3\sin(2000t)$ µC. (a) How much charge enters the terminal from $t = 0$ to $t = 0.5$ ms? (b) Find $i(t)$.

3. Let $i(t)$ shown in the following diagram flow through the circuit element in Fig. 3.5. With $i(t) = 0$ for $t < 0$, find the total charge at: (a) 1 s, (b) 2 s, (c) 3 s and (d) 4 s.

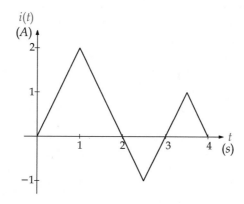

4. Let the charge entering the upper terminal in the circuit element in Fig. 3.5 be $q(t) = e^{-1000t}\sin(2000\pi t)$ C for $t \geq 0$. Determine the current for $t \geq 0$.

5. Find the power absorbed for the circuit element in Fig. 3.5 if (a) $v = 10$ V and $i = -2$ A, (b) $v = -10$ V and $i = -2$ A, (c) $v = -5$ V and $i = 2$ A, (d) $v = 10$ V and $i = 3$ A.

6. Find the power absorbed for the circuit element in Fig. 3.5 if (a) $v = 5$ V and $i = -2$ A, (b) $v = 5$ V and $i = 12$ A, (c) $v = -5$ V and $i = -5$ A, (d) $v = -5$ V and $i = 2$ A.

7. Find the power absorbed for the circuit element in Fig. 3.5 if

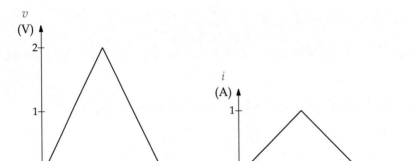

a.

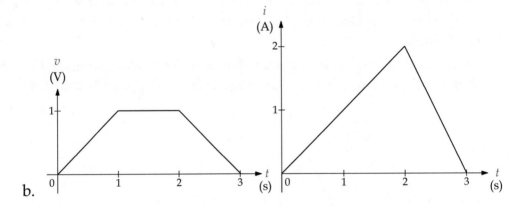

b.

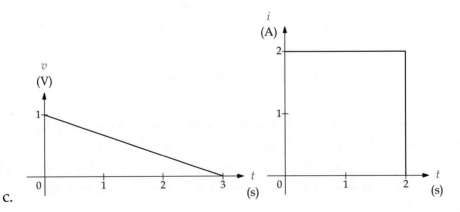

c.

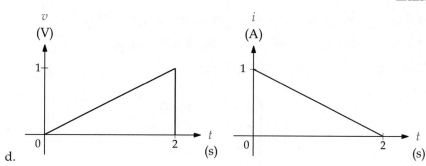

d.

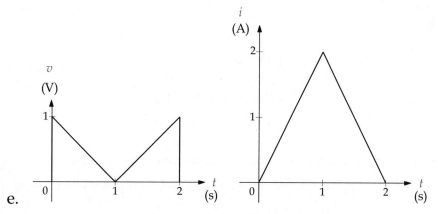

e.

8. Find the total energy delivered to the circuit element in Fig. 3.5 if

$$v = 3e^{-1000t}u(t)\,\text{V}$$

$$i = 5e^{-1000t}u(t)\,\text{A}$$

9. The voltage and current at the terminals in Fig. 3.5 are

$$v = e^{-500t}u(t)\,\text{V}$$
$$i = 2te^{-500t}u(t)\,\text{A}$$

(a) Find the time when the power is at its maximum.

(b) Find the energy delivered to the circuit element at $t = 0.004$ s.

(c) Find the total energy delivered to the circuit element.

10. The voltage and current at the terminals in Fig. 3.5 are

$$v = te^{-10,000t}u(t)\,\text{V}$$
$$i = (t + 10)e^{-10,000t}u(t)\,\text{A}$$

(a) Find the time when the power is at its maximum.

(b) Find the maximum power.

(c) Find the energy delivered to the circuit at $t = 1 \times 10^{-4}$ s.

(d) Find the total energy delivered to the circuit element.

11. The voltage at the terminals in Fig. 3.5 is

$$v = \begin{cases} 0\,V & t < 0 \\ t\,\text{V} & 0 \le t \le 1 \\ 2-t\,\text{V} & 1 < t \le 2 \\ 0\,\text{V} & t > 2 \end{cases}$$

If

$$p = \begin{cases} 0\,\text{W} & t < 0 \\ t^2\,\text{W} & 0 \le t \le 1 \\ t^2 - 4t + 4\,\text{W} & 1 < t \le 2 \\ 0\,\text{W} & t > 2 \end{cases}$$

how much charge enters the terminal between $t = 0$ and $t = 2$ s?

12. For the following circuit, find: (a) I_1, V_2 and V_3, (b) the power absorbed and delivered.

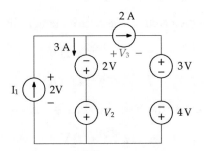

13. For the following circuit find (a) V_1, (b) I_2, (c) the power absorbed and delivered.

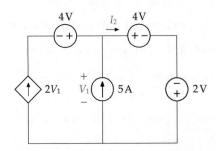

14. For the following circuit find (a) V_1, (b) the power absorbed and delivered.

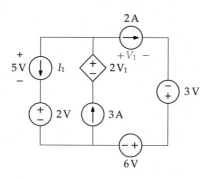

15. For the following circuit, find the power in each circuit element.

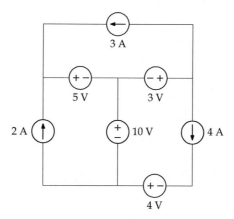

16. For the following circuit, find (a) I_1 and I_2, (b) power dissipated in each resistor, (c) show that the power dissipated equals the power generated.

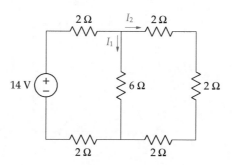

17. (a) Find the power dissipated in each resistor. (b) Show that the power dissipated equals the power generated.

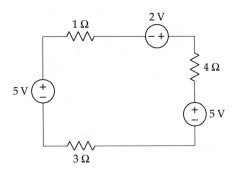

18. Find V_R in the following circuit.

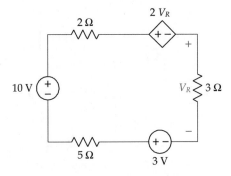

19. Find I in the following circuit.

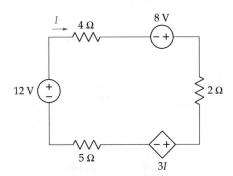

20. Find I_R in the following circuit.

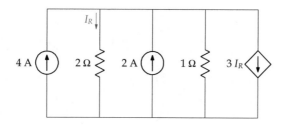

21. Find I_2 in the following circuit.

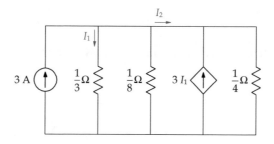

22. Find i_b for the following circuit.

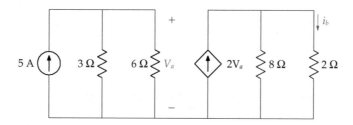

23. Find the equivalent resistance R_{ab} for the following circuit.

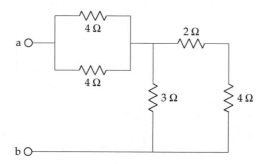

24. Find the equivalent resistance R_{ab} for the following circuit.

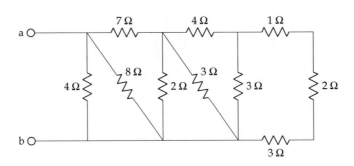

25. Find the equivalent resistance R_{ab} for the following circuit.

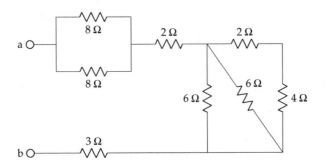

26. Find the equivalent resistance R_{ab} for the following circuit.

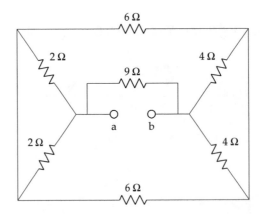

27. Find the equivalent resistance R_{ab} for the following circuit.

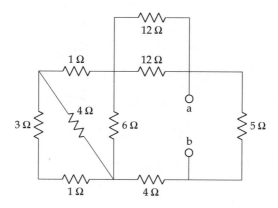

28. Find the equivalent resistance R_{ab} for the following circuit.

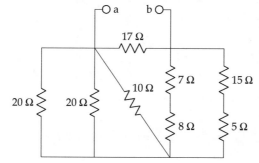

29. Find I_1 for the following circuit.

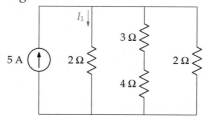

30. Find I_1 and V_1 for the following circuit.

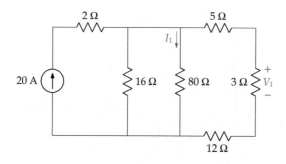

31. Find I_1 and V_1 for the following circuit.

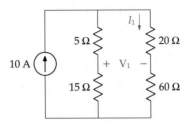

32. Find I_1, V_1 and V_2 for the following circuit.

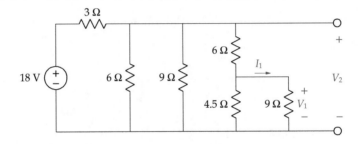

33. Use the node-voltage method to determine v_1 and v_2.

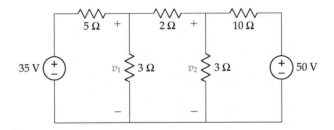

34. Use the node-voltage method to determine v_1 and v_2.

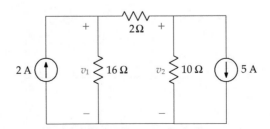

35. Use the node-voltage method to determine v_1 and v_2.

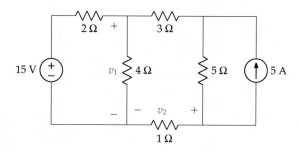

36. Use the node-voltage method to determine v_1 and v_2.

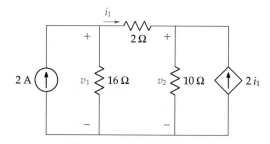

37. Use the node-voltage method to determine v_1 and v_2.

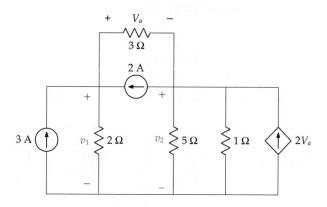

38. Use the node-voltage method to determine v_1 and v_2.

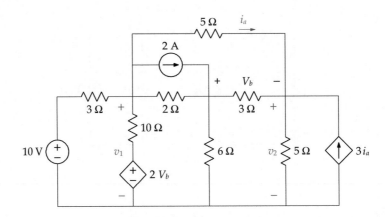

39. Use the node-voltage method to determine v_1 and v_2.

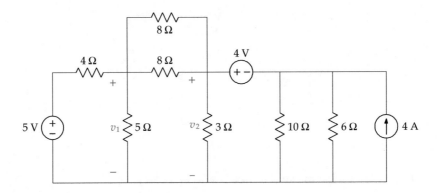

40. Use the node-voltage method to determine v_1 and v_2.

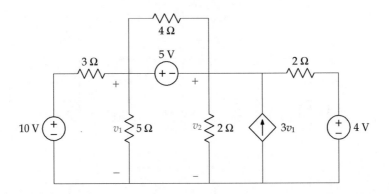

41. Use the node-voltage method to determine v_1 and v_2.

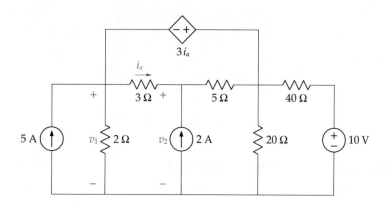

42. Use the node-voltage method to determine v_1 and v_2.

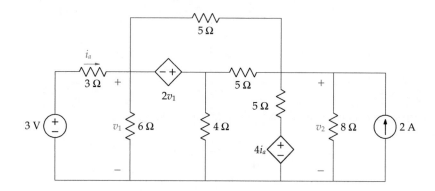

43. Use the node-voltage method to determine v_1 and v_2.

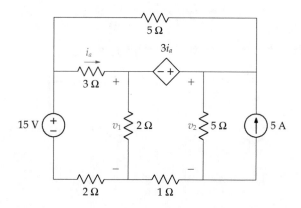

44.   Use the mesh-current method to determine i_1 and i_2.

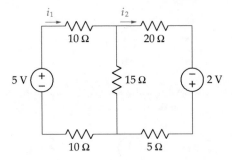

45.   Use the mesh-current method to determine i_1 and i_2.

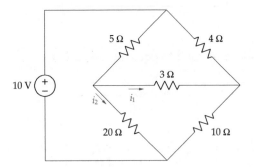

46.   Use the mesh-current method to determine i_1 and i_2.

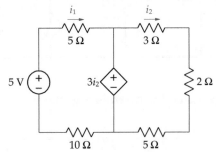

47.   Use the mesh-current method to determine i_1 and i_2.

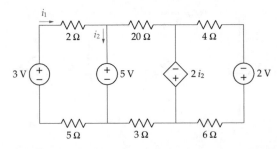

48. Use the mesh-current method to determine i_1 and i_2.

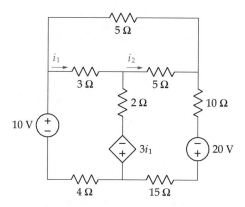

49. Use the mesh-current method to determine i_1 and i_2.

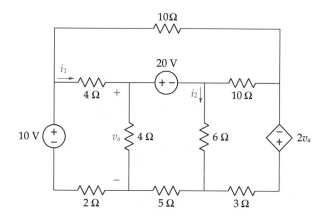

50. Use the mesh-current method to determine i_1 and i_2.

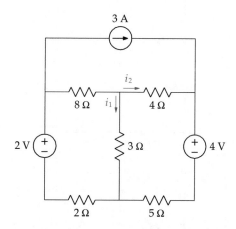

51. Use the mesh-current method to determine i_1 and i_2.

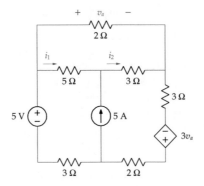

52. Use the mesh-current method to determine i_1 and i_2.

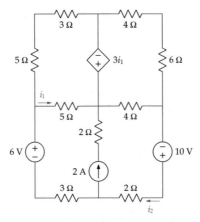

53. Use the mesh-current method to determine i_1 and i_2.

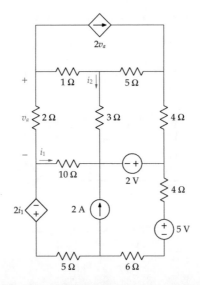

54. Use the mesh-current method to determine i_1 and i_2.

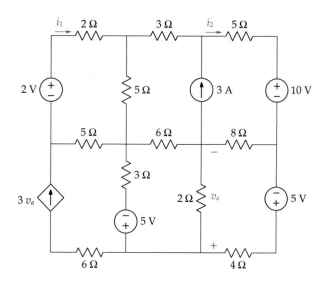

55. Use a series of source transformations and resistor combinations to find v_o.

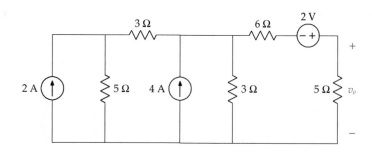

56. Use a series of source transformations and resistor combinations to find v_o.

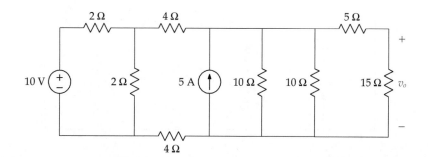

57.   Use a series of source transformations and resistor combinations to find v_o.

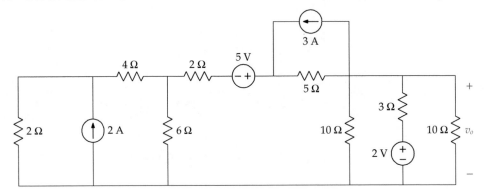

58.   Use the superposition method to find v_o.

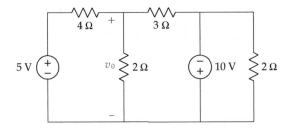

59.   Use the superposition method to find v_o.

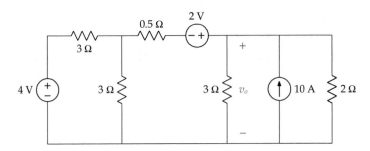

60.   Use the superposition method to find v_o.

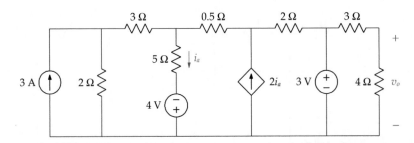

61. Use the superposition method to find v_o.

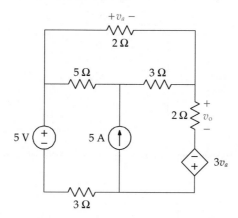

62. Find the Thévenin equivalent with respect to terminals a and b.

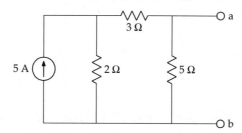

63. Find the Thévenin equivalent with respect to terminals a and b.

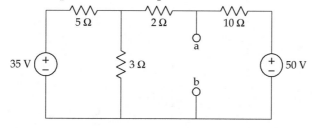

64. Find the Thévenin equivalent with respect to terminals a and b.

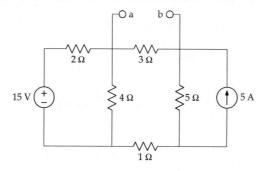

65. Find the Thévenin equivalent with respect to terminals a and b.

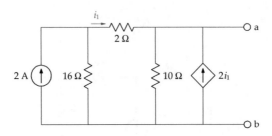

66. Find the Thévenin equivalent with respect to terminals a and b.

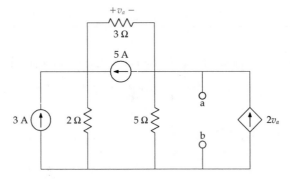

67. Find the Thévenin equivalent with respect to terminals a and b.

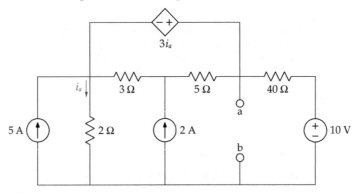

68. Find the Norton equivalent with respect to terminals a and b.

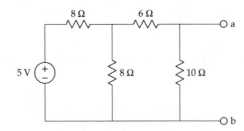

69. Find the Norton equivalent with respect to terminals a and b.

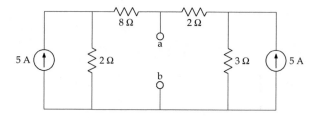

70. Find the Norton equivalent with respect to terminals a and b.

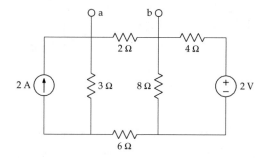

71. Find the Norton equivalent with respect to terminals a and b.

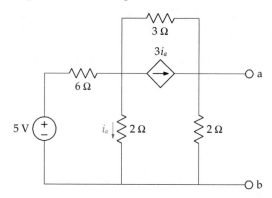

72. Find the Norton equivalent with respect to terminals a and b.

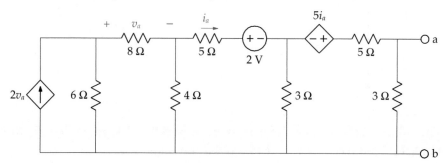

73. Find the Norton equivalent with respect to terminals a and b.

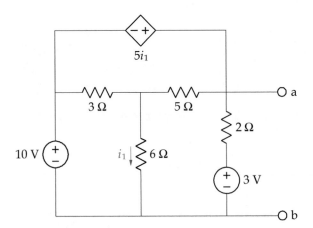

74. A current pulse given by $i(t) = (2 + 10e^{-2t})u(t)$ is applied through a 10 mH inductor. (a) Find the voltage across the inductor. (b) Sketch the current and voltage. (c) Find the power as a function of time.

75. A current pulse given by $i(t) = (5 + 3\sin(2t))u(t)$ is applied through a 2 mH inductor. Determine the voltage across the inductor.

76. The current pulse shown in the following figure is applied through a 5 mH inductor. Find the voltage, power and energy.

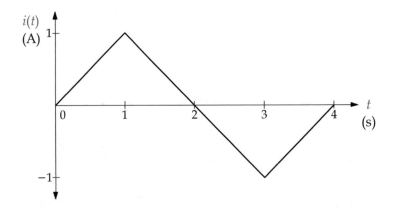

77. The voltage across an L = 2.5 mH inductor is $v(t) = 10\cos(1000t)$ mV, with $i(0) = 1$ mA. (a) Find $i(t)$ for $t \geq 0$. (b) Find the power and energy.

78. The voltage across an inductor is given by the following figure. If $L = 30$ mH and $i(0) = 0$ A, find $i(t)$ for $t \geq 0$.

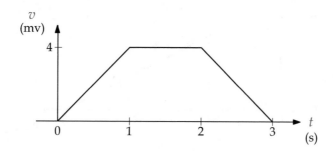

79. The voltage across an inductor is given by the following figure. If $L = 50$ mH and $i(0) = 0$ A, find $i(t)$ for $t \geq 0$.

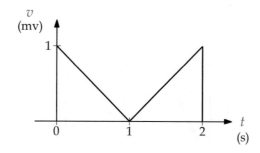

80. The voltage across a 4 µF capacitor is $v(t) = (200{,}000t - 50{,}000)e^{-2000t}u(t)$ V. Find (a) current through the capacitor, (b) power as a function of time, (c) energy.

81. The voltage across a 0.5 µF capacitor is $v(t) = (3 + 5e^{-2t})u(t)$ V. Find the current and power.

82. The voltage across a 1 µF capacitor is $v(t) = (5t + 3\sin(2t))e^{-3t}u(t)$ V. Find the current and power.

83. The current through a 5 µF capacitor is

$$i(t) = \begin{cases} 0 \text{ mA} & t < 0 \text{ ms} \\ 5t^2 \text{ mA} & 0 \leq t < 1 \text{ ms} \\ 5\left(2 - t^2\right) \text{ mA} & 1 < t \leq \sqrt{2} \text{ ms} \\ 0 \text{ mA} & t > \sqrt{2} \text{ ms} \end{cases}$$

Find the voltage across the capacitor.

84. The current through a 10 µF capacitor is given by the following figure. If $v(1) = 0$ V, find $v(t)$ for $t > 1$ s.

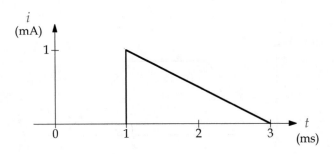

85. The current through a 100 µF capacitor is given by the following figure. If $v(0)$ is 0 V, find $v(t)$ for $t \geq 0$ s.

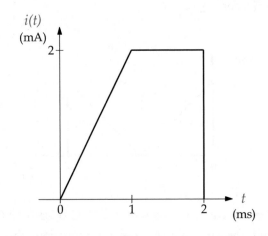

86. Find equivalent inductance between terminals a and b for the circuit in the following figure.

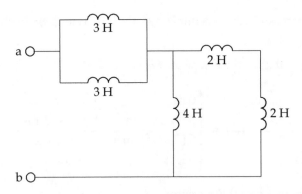

87. Find equivalent inductance between terminals a and b for the circuit in the following figure.

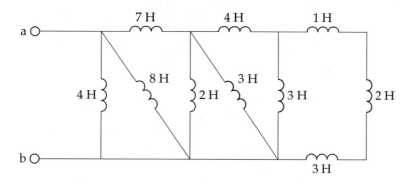

88. Find equivalent inductance between terminals a and b for the circuit in the following figure.

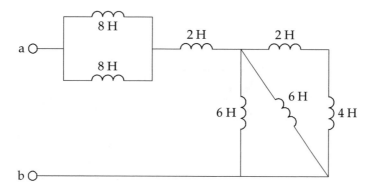

89. Find the equivalent capacitance between terminals a and b for the following circuit.

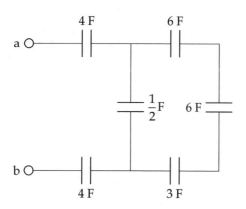

90. Find the equivalent capacitance between terminals a and b for the following circuit.

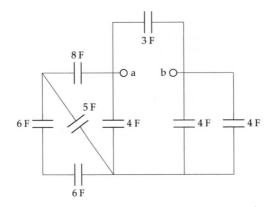

91. Find the equivalent capacitance between terminals a and b for the following circuit.

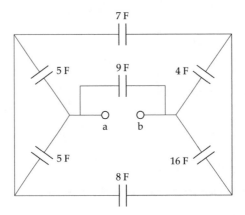

92. Reduce the following circuit to a single capacitor and inductor.

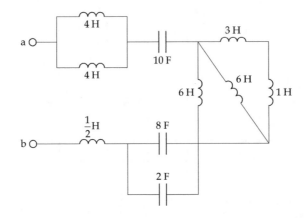

93. Reduce the following circuit to a single capacitor and inductor.

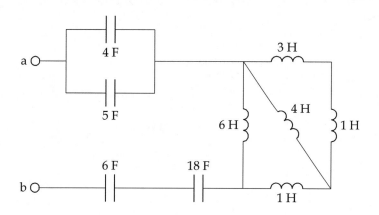

94. For the following circuit, $I_S = 2(1 - e^{-5t})u(t)$ A and $i(0) = 2$ A. (a) Find $v(t)$ for $t \geq 0$. (b) Find $i(t)$ for $t \geq 0$.

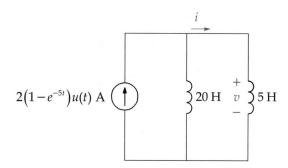

95. For the following circuit, $I_S = 5\sin 2t$ A and $i(0) = \frac{1}{2}$ A. (a) Find $v(t)$ for $t \geq 0$. (b) Find $i(t)$ for $t \geq 0$.

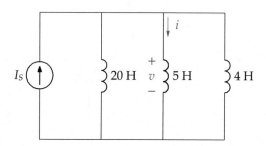

96. For the following circuit, $I_S = 5(1 - e^{-3t})\, u(t)$ A, $i_1(0) = 1$ A, and $i_2(0) = 2$ A. Find the following for $t \geq 0$: (a)$v_o(t)$, (b)$v_1(t)$, (c) $i_o(t)$, (d) $i_1(t)$, (e) $i_2(t)$.

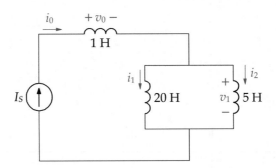

97. For the following circuit, $V_S = 3(1 - e^{-5t})\, u(t)$ V and $v(0) = 2$ V. Find the following for$t \geq 0$: (a)$i(t)$, (b) $v(t)$.

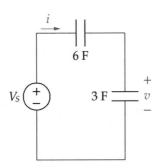

98. For the following circuit, $V_S = 5 \cos 3t$ V and $v(0) = 1$ V. Find the following for $t \geq 0$: (a) $i(t)$, (b) $v(t)$.

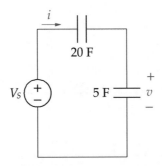

99. For the following circuit, $V_S = 2(1 - e^{-3t})\, u(t)$ V, $v_1(0) = 3$ V and $v_2(0) = 2$ V. Find
     the following for $t \geq 0$: (a) $i_1(t)$, (b) $v_1(t)$, (c) $v_2(t)$, (d) $i_2(t)$.

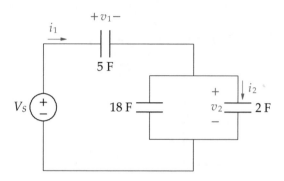

100. The switch has been in position a for a long time. At $t = 0$, the switch instantaneously
      moves to position b. Find i_L and v_1 for $t > 0$.

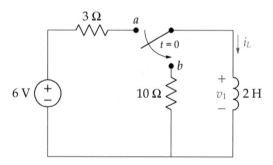

101. The switch has been in position a for a long time. At $t = 0$, the switch instantaneously
      moves to position b. Find i_L and v_1 for $t > 0$.

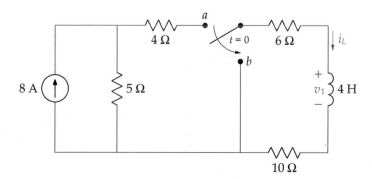

102. The switch has been in position *a* for a long time. At $t = 0$, the switch instantaneously moves to position *b*. Find i_L and v_1 for $t > 0$.

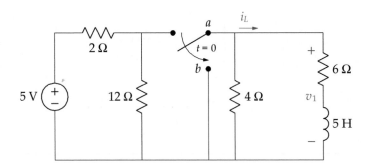

103. The switch has been in position *a* for a long time. At $t = 0$, the switch instantaneously moves to position *b*. Find i_L and v_1 for $t > 0$.

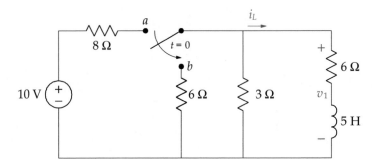

104. The switch has been in position *a* for a long time. At $t = 0$, the switch instantaneously moves to position *b*. Find i_L and v_1 for $t > 0$.

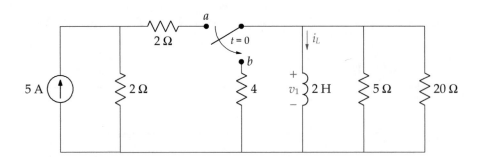

105. The switch has been in position a for a long time. At $t = 0$, the switch instantaneously moves to position b. Find i_L and v_1 for $t > 0$.

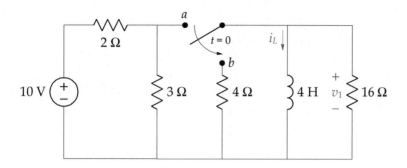

106. The switch has been in position a for a long time. At $t = 0$, the switch instantaneously moves to position b. Find i_L and v_1 for $t > 0$.

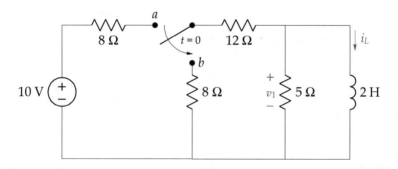

107. The switch has been in position a for a long time. At $t = 0$, the switch instantaneously moves to position b. Find i_L and v_1 for $t > 0$.

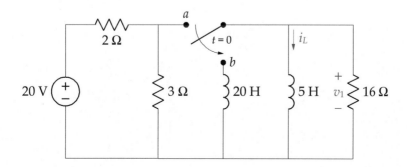

108. The switch has been in position a for a long time. At $t = 0$, the switch instantaneously moves to position b. Find i_L, i_1 and v_1 for $t > 0$.

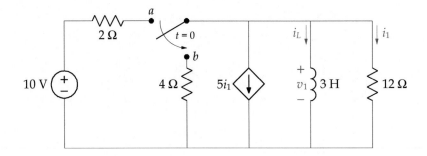

109. The switch has been in position a for a long time. At $t = 0$, the switch instantaneously moves to position b. Find i_L, i_1 and v_1 for $t > 0$.

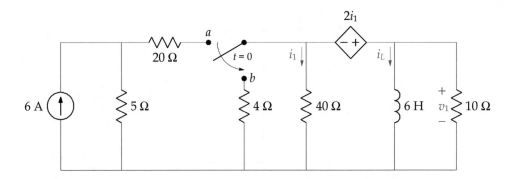

110. The switch has been in position a for a long time. At $t = 0$, the switch instantaneously moves to position b. Find v_c and i_1 for $t > 0$.

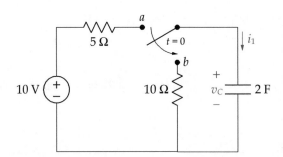

111. The switch has been in position *a* for a long time. At $t = 0$, the switch instantaneously moves to position *b*. Find v_c and i_1 for $t > 0$.

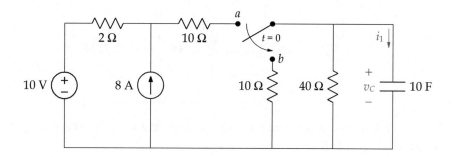

112. The switch has been in position *a* for a long time. At $t = 0$, the switch instantaneously moves to position *b*. Find v_c and i_1 for $t > 0$.

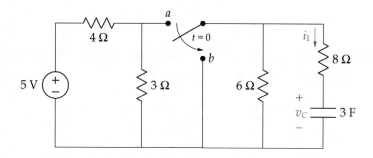

113. The switch has been in position *a* for a long time. At $t = 0$, the switch instantaneously moves to position *b*. Find v_c and i_1 for $t > 0$.

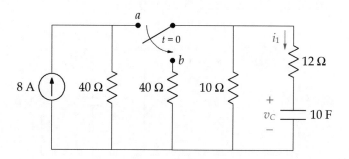

114. The switch has been in position *a* for a long time. At $t = 0$, the switch instantaneously moves to position *b*. Find v_c and i_1 for $t > 0$.

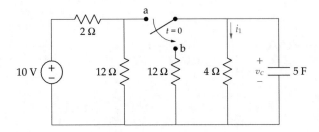

115. The switch has been in position *a* for a long time. At $t = 0$, the switch instantaneously moves to position *b*. Find v_c and i_1 for $t > 0$.

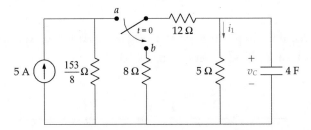

116. The switch has been in position *a* for a long time. At $t = 0$, the switch instantaneously moves to position *b*. Find v_c and i_1 for $t > 0$.

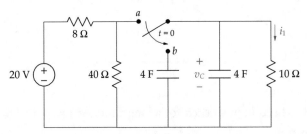

117. The switch has been in position *a* for a long time. At $t = 0$, the switch instantaneously moves to position *b*. Find v_c and i_1 for $t > 0$.

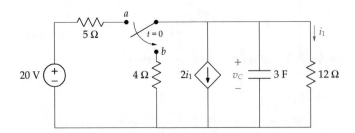

118. The switch has been in position *a* for a long time. At $t = 0$, the switch instantaneously moves to position *b*. Find v_c and i_1 for $t > 0$.

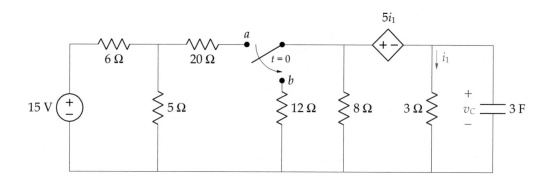

119. The switch has been in position *a* for a very long time. At $t = 0$, the switch moves to position *b*. Find i_L and v_1 for $t > 0$.

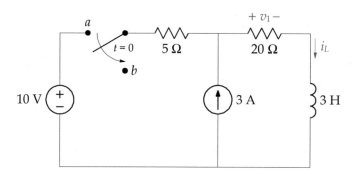

120. The switch has been in position *a* for a long time. At $t = 0$, the switch instantaneously moves to position *b*. Find i_L and v_1 for $t > 0$.

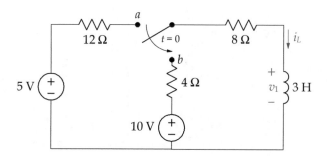

121. The switch has been in position a for a long time. At $t = 0$, the switch instantaneously moves to position b. Find i_L and v_1 for $t > 0$.

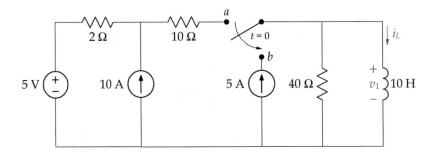

122. The switch has been in position a for a long time. At $t = 0$, the switch instantaneously moves to position b. Find i_L and v_1 for $t > 0$.

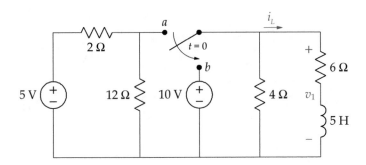

123. The switch has been in position a for a long time. At $t = 0$, the switch instantaneously moves to position b. Find i_1 and v_1 for $t > 0$.

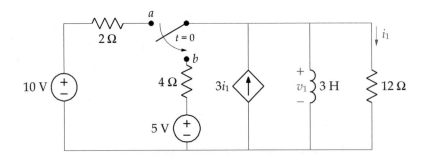

124. The switch has been in position *a* for a long time. At $t = 0$, the switch instantaneously moves to position *b*. Find i_L, i_1 and v_1 for $t > 0$.

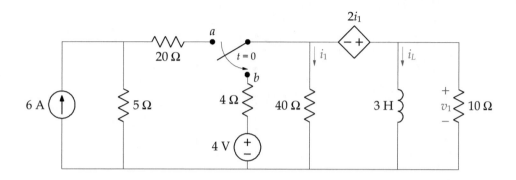

125. The switch has been position *a* for a very long time. At $t = 0$, the switch closes. Find v_C and i_1 for $t > 0$.

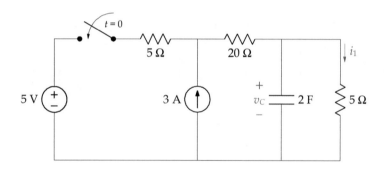

126. The switch has been in position *a* for a long time. At $t = 0$, the switch instantaneously moves to position *b*. Find v_c and i_1 for $t > 0$.

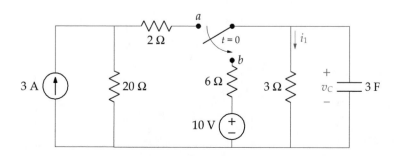

127. The switch has been in position a for a long time. At $t = 0$, the switch instantaneously moves to position b. Find v_c and i_1 for $t > 0$.

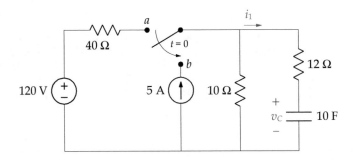

128. The switch has been in position a for a long time. At $t = 0$, the switch instantaneously moves to position b. Find v_c and i_1 for $t > 0$.

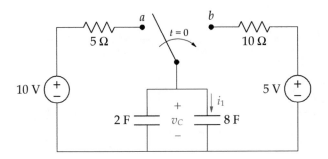

129. The switch has been in position a for a long time. At $t = 0$, the switch instantaneously moves to position b. Find v_c and i_1 for $t > 0$.

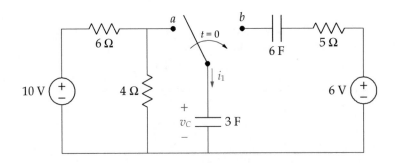

130. The switches have been in positions *a* and *c* for a long time. At $t = 0$, switch *ab* instantaneously moves to position *b*. At $t = 0.06$, switch *cd* instantaneously moves to position *d*. Find i_1 and v_c for $t > 0$.

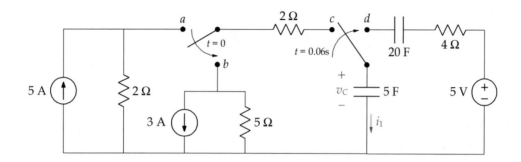

131. The switches have been in positions *a* and *c* for a long time. At $t = 0$, switch *ab* instantaneously moves to position *b*. At $t = 0.15$, switch *cd* instantaneously moves to position *d*. Find i_1 and v_c for $t > 0$.

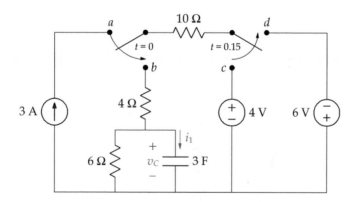

132. Find i_L and v_c for $t > 0$ for the following circuit if: (a) $i_s = 3u(t)$ A; (b) $i_s = 1 + 3u(t)$ A.

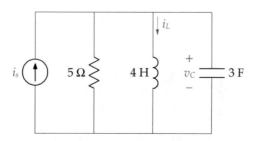

133. Find i_L and v_c for $t > 0$ for the following circuit if: (a) $v_s = 5u(t)$ V; (b) $v_s = 2 + 5u(t)$ V.

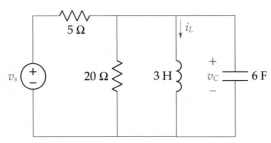

134. Find i_L and v_c for $t > 0$ for the following circuit if: (a) $i_s = 10u(t)$ A; (b) $i_s = -1 + 10u(t)$ A.

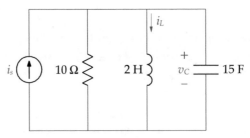

135. Find i_L and v_c for $t > 0$ for the following circuit if: (a) $v_s = 20u(t)$ V; (b) $v_s = 4 + 20u(t)$ V.

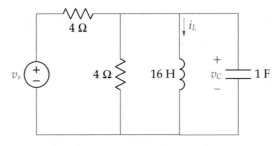

136. Find i_L and v_c for $t > 0$ for the following circuit if: (a) $i_s = 3u(t)$ A; (b) $i_s = -1 + 3u(t)$ A.

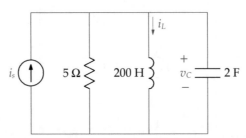

137. Find i_L and v_c for $t > 0$ for the following circuit if: (a) $v_s = 5u(t)$ V; (b) $v_s = 12 + 5u(t)$ V.

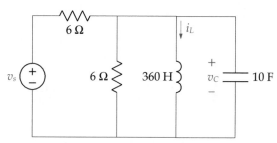

138. Find i_L and v_c for $t > 0$ for the following circuit if: (a) $i_s = 5u(t)$ A; (b) $i_s = 5 + 5u(t)$ A.

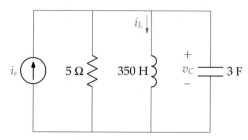

139. Find i_L and v_c for $t > 0$ for the following circuit if: (a) $v_s = 3u(t)$ V; (b) $v_s = 2 + 3u(t)$ V.

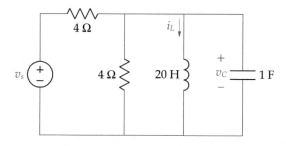

140. Find i_L and v_c for $t > 0$ for the following circuit if: (a) $i_s = 10u(t)$ A; (b) $i_s = -1 + 10u(t)$ A.

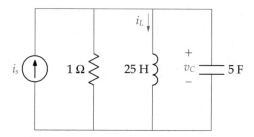

141. Find i_L and v_c for $t > 0$ for the following circuit if: (a) $v_s = 5u(t)\,\text{V}$; (b) $v_s = 5u(t) + 3\,\text{V}$.

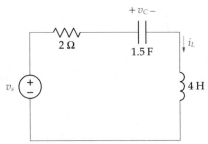

142. Find i_L and v_c for $t > 0$ for the following circuit if: (a) $i_s = 3u(t)\,\text{A}$; (b) $i_s = 3u(t) - 1\,\text{A}$.

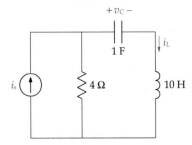

143. Find i_L and v_c for $t > 0$ for the following circuit if: (a) $i_s = 10u(t)\,\text{A}$; (b) $i_s = 10u(t) + 5\,\text{A}$.

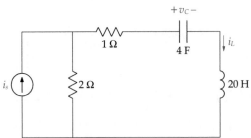

144. Find i_L, i_1 and v_c for $t > 0$ for the following circuit if: (a) $v_s = 5u(t)\,\text{V}$; (b) $v_s = 5u(t) + 2\,\text{V}$.

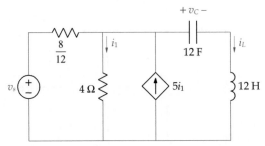

145. Find i_L and v_c for $t > 0$ for the following circuit if: (a) $i_s = 6u(t)$ A; (b) $i_s = 3u(t) + 1$ A.

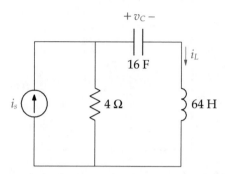

146. Find i_L and v_c for $t > 0$ for the following circuit if: (a) $v_s = 4u(t)$ V; (b) $v_s = 4u(t) - 2$ V.

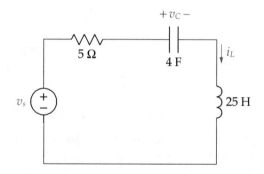

147. Find i_L and v_c for $t > 0$ for the following circuit if: (a) $v_s = 2u(t)$ V; (b) $v_c = 2u(t) + 2$ V.

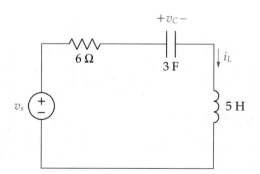

148. Find i_L and v_c for $t > 0$ for the following circuit if: (a) $i_s = 2u(t)$ A; (b) $i_s = 2u(t) - 1$ A.

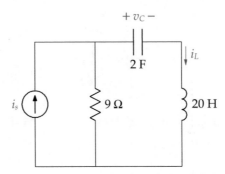

149. Find i_L and v_c for $t > 0$ for the following circuit if: (a) $v_s = 5u(t)$ V; (b) $v_s = 5u(t) - 2$ V.

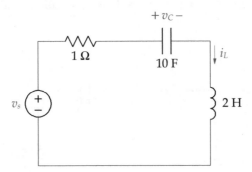

150. For the following circuit we are given that $i_{L_1}(0) = 2$ A, $i_{L_2}(0) = 5$ A, $v_{c_1}(0) = 2$ V, $v_{c_2}(0) = -3$ V and $i_s = 2e^{-2t}u(t)$ A. (a) Write the node-voltage equations necessary to solve this circuit. (b) Write the mesh-current equations necessary to solve this circuit. (c) Use the node-voltage method to find v_b for $t > 0$. (d) Use the mesh-current method to find v_b for $t > 0$.

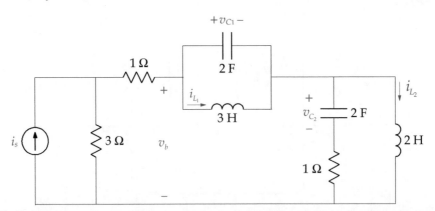

151. Find v_{c_1} for $t > 0$ for the following circuit: (a) using the node-voltage method if $v_s = 2e^{-3t}u(t)$ V; (b) using the mesh-current method if $v_s = 2e^{-3t}u(t)$ V; (c) using the node-voltage method if $v_s = 3\cos(2t)\,u(t)$ V; (d) using the mesh-current method if $v_s = 3\cos(2t)\,u(t)$ V; (e) using the node-voltage method if $v_s = 3u(t) - 1$ V; (f) using the mesh-current method if $v_s = 3u(t) - 1$ V.

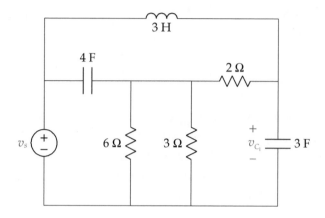

152. Find i_{L_1} for $t > 0$ for the following circuit: (a) using the node-voltage method if $i_s = 2e^{-3t}u(t)$ A; (b) using the mesh-current method if $i_s = 2e^{-3t}u(t)$ A; (c) using the node-voltage method if $i_s = 3\cos(2t)\,u(t)$ A; (d) using the mesh-current method if $i_s = 3\cos(2t)\,u(t)$ A; (e) using the node-voltage method if $i_s = 2u(t) + 2$ A; (f) using the mesh-current method if $i_s = 2u(t) + 2$ A.

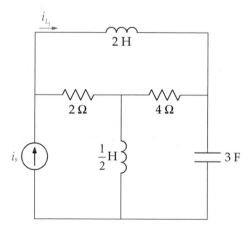

153. Find v_c for $t > 0$ for the following circuit: (a) using the node-voltage method if $v_s = 3e^{-5t}u(t)$ V; (b) using the mesh-current method if $v_s = 3e^{-5t}u(t)$ V; (c) using the node-voltage method if $v_s = 3\sin(5t)\,u(t)$ V; (d) using the mesh-current method if

$v_s = 3\sin(5t)\,u(t)\,$V; (e) using the node-voltage method if $v_s = 5u(t) - 2\,$V; (f) using the mesh-current method if $v_s = 5u(t) - 2\,$V.

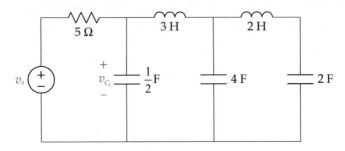

154. Find i_L for $t > 0$ for the following circuit: (a) using the node-voltage method if $i_s = 5e^{-3t}u(t)\,$A; (b) using the mesh-current method if $i_s = 5e^{-3t}u(t)\,$A; (c) using the node-voltage method if $i_s = 2\sin(3t)\,u(t)\,$A; (d) using the mesh-current method if $i_s = 2\sin(3t)\,u(t)\,$A; (e) using the node-voltage method if $i_s = 2u(t) + 3\,$A; (f) using the mesh-current method if $i_s = 2u(t) + 3\,$A.

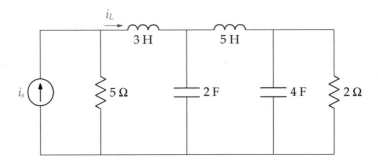

155. The switch has been in position a for a long time. At $t = 0$, the switch moves to position b, then at $t = 4$, the switch moves back to position a. Find v_c for $t > 0$.

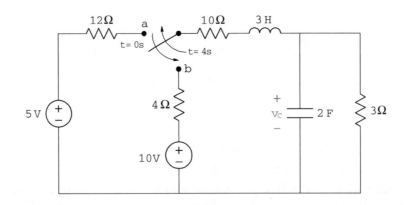

156. The switches operate as indicated in the following circuit. Find v_c for $t > 0$.

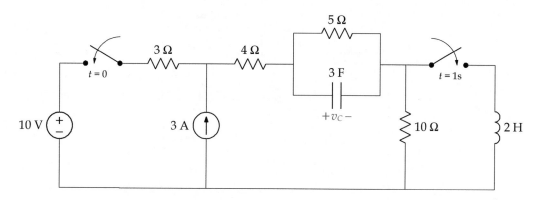

157. The switch has been in position a for a long time. At $t = 0$, the switch moves to position b, and then back to position a at $t = 2$ s. Find v_c for $t > 0$.

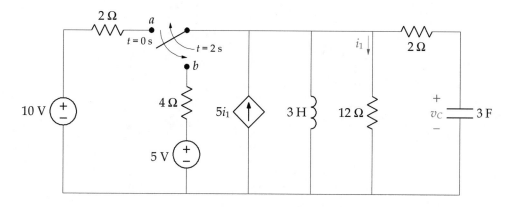

158. The switches operate as indicated in the following circuit. Find i_L for $t > 0$.

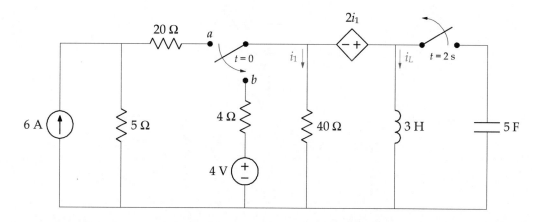

159.   The operational amplifier shown in the following figure is ideal. Find v_o and i_o.

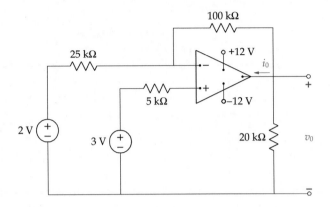

160.   The operational amplifier shown in the following figure is ideal. Find v_o and i_o.

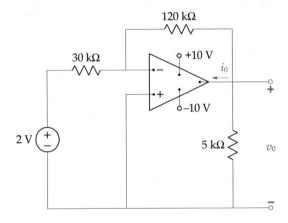

161.   The operational amplifier shown in the following figure is ideal. Find v_o and i_o.

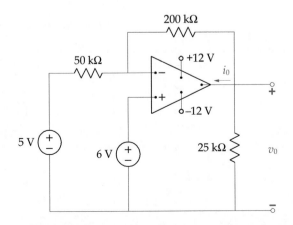

162. The operational amplifier shown in the following figure is ideal. Find v_o.

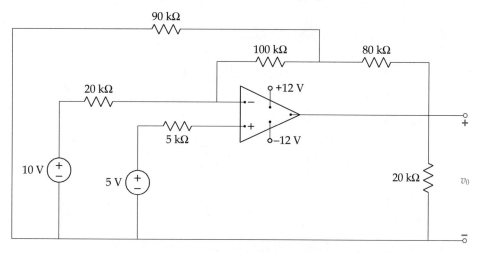

163. Find the overall gain for the following circuit if the operational amplifier is ideal. Draw a graph of v_o versus V_s if V_s varies between 0 and 10 V.

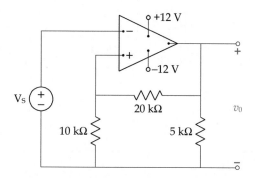

164. Find v_o in the following circuit if the operational amplifier is ideal.

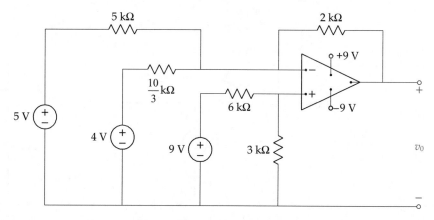

165. Find v_o in the following circuit if the operational amplifier is ideal.

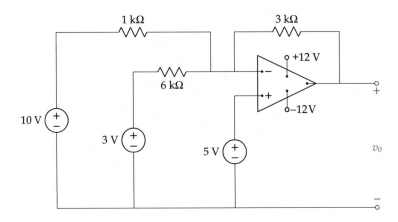

166. Find i_o in the following circuit if the operational amplifiers are ideal.

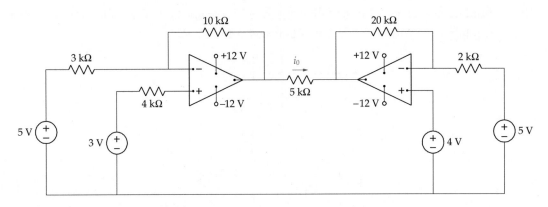

167. Find v_o in the following circuit if the operational amplifiers are ideal.

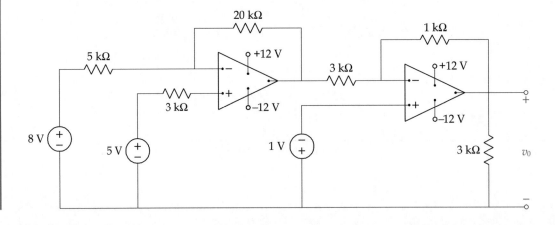

168. Suppose the input V_s is given as a triangular waveform as shown in the following figure. If there is no stored energy in the following circuit with an ideal operational amplifier, find v_o.

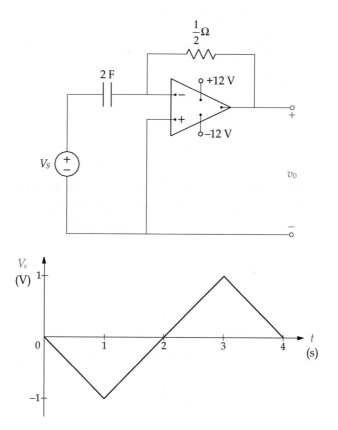

169. Suppose the input V_s is given in the following figure. If there is no stored energy in the following circuit with an ideal operational amplifier, find v_o.

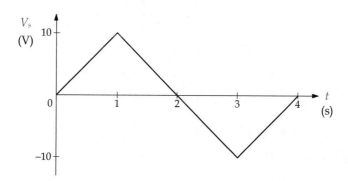

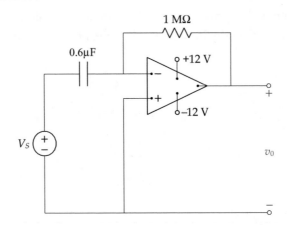

170. Suppose the input V_s is given in the following figure. If there is no stored energy in the following circuit with an ideal operational amplifier, find v_o.

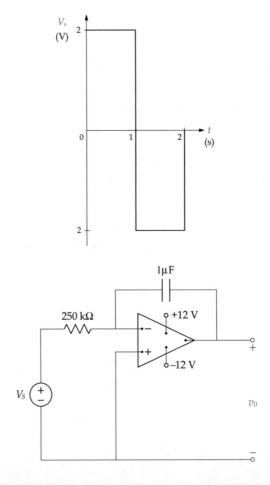

171. Suppose the input V_s is given in the following figure. If there is no stored energy in the following circuit with an ideal operational amplifier, find v_o.

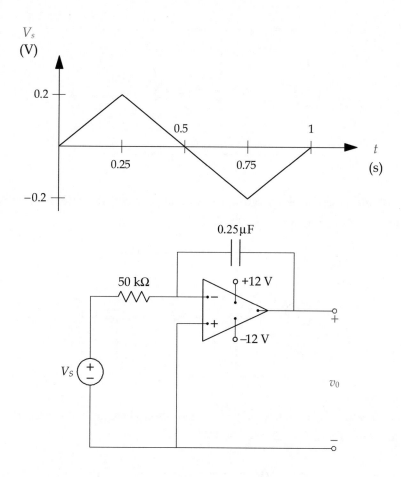

172. The following circuit is operating in the sinusoidal steady state. Find the steady-state expression for i_L if $i_s = 30 \cos 20t$ A.

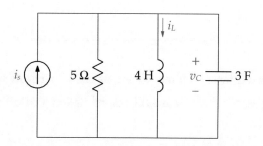

173. The following circuit is operating in the sinusoidal steady state. Find the steady-state expression for v_c if $v_s = 10 \sin 1000t$ V.

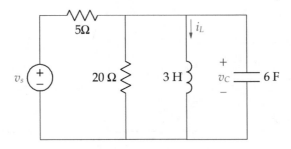

174. The following circuit is operating in the sinusoidal steady state. Find the steady-state expression for i_L if $i_s = 5 \cos 500t$ A.

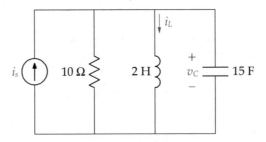

175. The following circuit is operating in the sinusoidal steady state. Find the steady-state expression for v_c if $i_s = 25 \cos 4000t$ V.

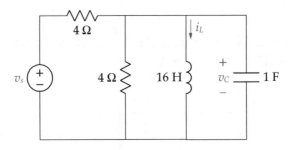

176. Design a low-pass filter with a magnitude of 10 and a cutoff frequency of 250 rad/s.

177. Design a high-pass filter with a magnitude of 20 and a cutoff frequency of 300 rad/s.

178. Design a band-pass filter a gain of 15 and pass through frequencies from 50 to 200 rad/s.

179. Design a low-pass filter with a magnitude of 5 and a cutoff frequency of 200 rad/s.

180. Design a high-pass filter with a magnitude of 10 and a cutoff frequency of 500 rad/s.

181. Design a band-pass filter a gain of 10 and pass through frequencies from 20 to 100 rad/s.

182. Suppose the operational amplifier in the following circuit is ideal. (The circuit is a low-pass first-order Butterworth filter.) Find the magnitude of the output v_o as a function of frequency.

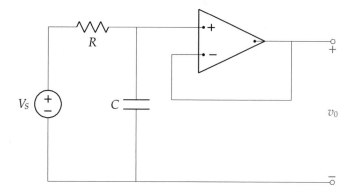

183. With an ideal operational amplifier, the following circuit is a second-order Butterworth low-pass filter. Find the magnitude of the output v_o as a function of frequency.

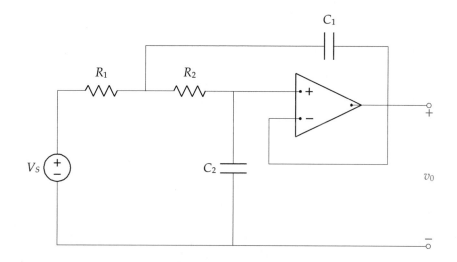

184. A third-order Butterworth low-pass filter is shown in the following circuit with an ideal operational amplifier. Find the magnitude of the output v_o as a function of frequency.

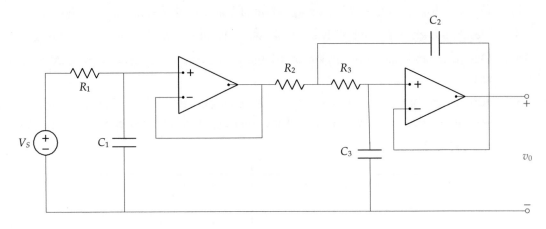

Printed in the United States
by Baker & Taylor Publisher Services